GENEVE

Begin your own tradition.

You never actually own
a Patek Philippe.

You merely take care of it for
the next generation.

Calatrava Ref. 6006G

de BOULLE

6821 Preston Road
Dallas, TX 75205
214-522-2400

River Oaks District
Houston, TX 77027
713-621-2400

deBoulle.com

EST 1951

RAZNY
JEWELERS

CHICAGO . ADDISON . HIGHLAND PARK
RAZNY.COM

WATCHES INTERNATIONAL

THE ORIGINAL ANNUAL OF THE WORLD'S FINEST WRISTWATCHES

First published in the United States in 2019 by

TOURBILLON INTERNATIONAL
A MODERN LUXURY COMPANY

11 West 25th Street, 8th Floor
New York, NY 10010
Tel: +1 (212) 627-7732

Caroline Childers
PUBLISHER

Michel Jeannot
EDITOR IN CHIEF

Lew Dickey
CHAIRMAN

Michael Dickey
CHIEF EXECUTIVE OFFICER

John Dickey
CHIEF OPERATING OFFICER

Isabella Lee
SENIOR COUNSEL

In association with **RIZZOLI** INTERNATIONAL PUBLICATIONS, INC.

300 Park Avenue South, New York, NY 10010

COPYRIGHT©2019 TOURBILLON INTERNATIONAL, ALL RIGHTS RESERVED

No part of this publication may be reproduced in any manner whatsoever without prior written permission from Tourbillon International.

ISBN: 978-0-8478-6670-0

DISCLAIMER: THE INFORMATION CONTAINED IN WATCHES INTERNATIONAL VOL. XX HAS BEEN PROVIDED BY THIRD PARTIES. WHILE WE BELIEVE THESE SOURCES TO BE RELIABLE, WE ASSUME NO RESPONSIBILITY OR LIABILITY FOR THE ACCURACY OF TECHNICAL DETAILS CONTAINED IN THIS BOOK.

EVERY EFFORT HAS BEEN MADE TO LOCATE THE COPYRIGHT HOLDERS OF MATERIALS PRINTED IN THIS BOOK. SHOULD THERE BE ANY ERRORS OR OMISSIONS, WE APOLOGIZE AND SHALL BE PLEASED TO MAKE ACKNOWLEDGMENTS IN FUTURE EDITIONS.

PRINTED IN ITALY

COVER: RICHARD ORLINSKI AND HUBLOT CLASSIC FUSION TOURBILLON SAPPHIRE ORLINSKI

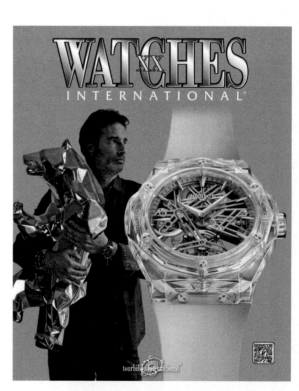

Cartier

PANTHÈRE DE CARTIER COLLECTION

BVLGARI.COM   #LIFEISNOTROUND

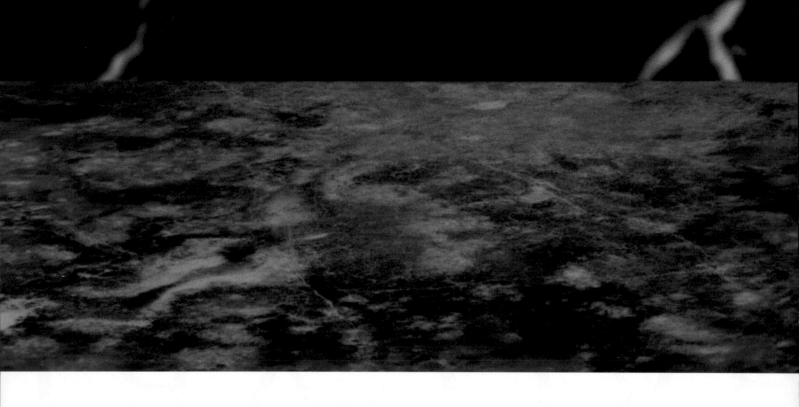

SERPENTI
BVLGARI.COM

[Letter from the President]

Race for Precision

The world of watchmaking has been in the constant pursuit of time measuring precision since the beginning of mechanical watchmaking in the 14th century, mirroring and influencing the evolution of human civilization itself. Although the notion of time precision has evolved at a sometimes dizzying pace, the mechanical watch has maintained its relevance in the contemporary world.

In an age where time is broken down in increments of thousandths of a second, and we seem to have reached the boundary of precision, watchmakers surprise us every year by surpassing themselves, and each other.

In keeping with the *Watches International* tradition, we celebrate both history and innovation by looking at how far we have come and asking: what is next?

From the early time measurement developments created by the Mesopotamians, ancient Greeks and Aristotle, to the birth of mechanical clocks during the High Middle Ages and the subsequent chase to perfect them, we examine the role watches have played in society throughout history, as well as take a closer look at the fascinating competition between watchmakers who defy expectations and outdo themselves in their never-ending efforts to keep us, watch lovers, captivated.

In its quest for precision, watchmaking has always found inspiration and support in sports, another high-performance universe. This alliance also traces back to the birth of both worlds (Huygens, the founder of scientific watchmaking in the 17th century, was motivated in his investigations by the needs of sailors who were conquering the seas and discovering new worlds), and is very much alive today, with brands like TAG Heuer, Rolex, and Breitling, among others, offering regatta-themed models.

A healthy competition between countries on all continents also emerged as a result of researches for time precision, with Switzerland, the traditional land of watchmaking, trying to remain at the forefront of this craft, while innovations in the field have been made in America, the U.K. and Japan.

No matter what happens next, the ambitious watchmaking industry is set to continue to surpass itself, to the delight of its passionate, sophisticated and educated consumers.

Michael Dickey

New York 48 East 57 Street, New York, New York +1.212.719.5887
Geneva Chemin de Plein-Vent 1 ch-1228 Geneva, Switzerland +41.22.310.6962

jacobandco.com

[**Letter from the Publisher**]

The thousand and one lives of a watch

Is it an effect of the crisis? A regressive need to rediscover the imagined happiness of childhood? A strange vintage fever has seized watchmaking. It started a few years ago, with the desire to break with a hyper-consumerist society. Then an unfathomable nostalgia engulfed everything in its path. It seems like every watch brand is celebrating the anniversary of an iconic piece, or commemorating an event by re-releasing a model or collection… need we mention that in 2019 we will salute the 50th anniversary of the first step on the moon?

Even the used watch market has become democratized. It is now possible to resell or purchase authenticated and certified pre-owned watches on the internet. Are you looking for a TAG Heuer chronograph, a 1980s Rolex, the Breitling Emergency equipped with a two-frequency distress beacon, or even a Richard Mille Felipe Massa that costs more than $100,000? Watchfinder, Watchbox or Bucherer, to name just a few examples, fulfill your dream of wearing a timepiece with rich history and meaning, via a huge virtual "flea market."

This practice has been embraced by the industry. Surfing the retro wave, paradoxically, allows brands to lay claim to the future (and quite a fruitful future it is). Sixteen years after its creation, Watchfinder—the site that has become the specialist in high-end private second-hand sales—was bought by Richemont in June 2018. With this acquisition, which follows that of Yoox Net-à-Porter (YNAP), Johann Rupert's group is going on the offensive to cover new sectors, rejuvenate its image and stake a claim to a niche market that promises to grow exponentially. The other major players are also positioning themselves in the niche by developing dedicated online platforms, or even physical shops exclusively reserved for vintage watches. Everyone is targeting a customer disappointed by today's frantic consumerism, one who swears by the charm of the old.

If you come across a trendy hipster in a vintage checkered shirt, or a hip, artsy girl in a fifties pencil skirt, you might presume that their watch is a vintage model as well. A memory of the past? No, a new talisman.

Caroline Childers

Breguet
Depuis 1775

Breguet La Marine
Chronograph 5527

The New Tambour Horizon
Our journey, connected.

THE ORIGINAL ANNUAL OF THE WORLD'S FINEST WRISTWATCHES

TOURBILLON INTERNATIONAL
A MODERN LUXURY COMPANY
ADMINISTRATION, ADVERTISING SALES, EDITORIAL

11 West 25th Street, 8th Floor • New York, NY 10010
Tel: +1 (212) 627-7732

Caroline Childers
PUBLISHER

Michel Jeannot
EDITOR IN CHIEF

EDITORS	Elise Nussbaum
	Doris Sangeorzan
ART DIRECTOR	Mutsumi Hyuga
JUNIOR ASSISTANT EDITOR	Amber Ruiz
TRANSLATIONS	Susan Jacquet
VICE PRESIDENT OF OPERATIONS	Sean Bertram
INTERNATIONAL DIGITAL MEDIA DIRECTOR	Caroline Elbaz
INTERNATIONAL DIGITAL COORDINATOR	Eric Jean-Bart
PRODUCTION DIRECTOR	Alexandra Knerly
DIRECTOR OF INFORMATION TECHNOLOGY	Scott Brookman
NATIONAL CIRCULATION MANAGER	Maria Blondeaux
NATIONAL DISTRIBUTION MANAGER	Hector Galvez

PHOTOGRAPHY
Photographic Archives
Property of Tourbillon International, a Modern Luxury Company

MODERN **LUXURY**

Lew Dickey
CHAIRMAN

Michael Dickey
CHIEF EXECUTIVE OFFICER

CHIEF OPERATING OFFICER	John Dickey
SENIOR COUNSEL	Isabella Lee

COLLECTION

Fifty Fathoms

RAISE AWARENESS,
TRANSMIT OUR PASSION,
HELP PROTECT THE OCEAN

www.blancpain-ocean-commitment.com

BLANCPAIN

MANUFACTURE DE HAUTE HORLOGERIE

[Letter from the Editor in Chief]

Better to be wealthy and healthy...

We often discuss watchmaking as a perfectly coherent and united whole. We try to draw general conclusions, to unearth global trends, to describe a single reality—in short, we grasp for a simple and indisputable explanation. But the multi-faceted nature of reality catches us up short: explanations slide away, general theories cannot be applied to the whole, and oversimplifications are complicated by analysis of the facts. Depending on where we look, the realities are very different.

The results of watchmaking in 2018 are a new demonstration of this. With more than 6% growth, Swiss watch exports have reached a nice cruising speed. Overall, the situation seems indisputable, but the realities on the ground are very diverse.

Large brands have generally benefited more from the recovery, now entering its third year, than small entities. The brands most invested in China have, of course, benefited more than others from the area's strong recovery in demand in the first half of 2018. The most active companies in Asia have even benefited more from the growing godsend of Chinese tourists—and spenders!—worldwide. Active brands in the high-end sectors of the market fared better than entry-level brands, likely impacted by the arrival of smartwatches. In addition, companies that took on the digital shift early and are active on all digital channels show a better outlook than those who still believe that e-commerce is not suitable for selling watches. Finally, the results of luxury groups clearly show that jewelry is much more dynamic than watchmaking. These memorable mood swings affect the entire industry.

In 2018, then, we can say that it was advantageous to be a large brand, also active in jewelry, ideally backed by a group, very active in Asia, very connected, and rather upscale, than to be a watch brand dependent on companies selling its timepieces mainly in Europe to a local clientele... And again, there are always exceptions! And what will happen in 2019? If no one dares to risk prognosis, one certainty remains: it will always be better to be rich and healthy than poor and sick. That truism holds true for watchmaking as well.

Michel Jeannot

LIMELIGHT GALA

arije

"Là, tout n'est qu'ordre et beauté,

Luxe, calme et volupté."

Charles Baudelaire

50, rue Pierre Charron – 75008 Paris

+33 1 47 20 50 50

5, place du Québec – 75006 Paris

+ 33 1 43 35 05 05

3, rue de Castiglione – 75001 Paris

+ 33 1 42 60 37 77

207A Sloane Street – SW1X9QX London

+44 20 7823 2852

[Summary]

8	Letter from the President
12	Letter from the Publisher
18	Letter from the Editor in Chief
22	Arije Profile
36	Westime Profile
40	Interview with Mr. Greg Simonian, President of Westime
44	Web Site Directory
48	Interview with Mr. Georges Kern, CEO of Breitling
54	Interview with Mr. Ricardo Guadalupe, CEO of Hublot
64	Interview with Mr. Walter von Känel, CEO of Longines
70	Interview with Mr. Richard Mille, CEO of Richard Mille
76	Interview with Mr. Julien Tornare, CEO of Zenith
82	Precisely Speaking, by Michel Jeannot
178	Brand Profiles and Watch Collections
384	Brand Directory

We assemble every single watch twice.
Because perfection takes time.

For us, perfection is a matter of principle. This is why, on principle, we craft all timepieces with the same care and assemble each watch twice. Thus, after the Lange 1 Moon Phase has been assembled for the first time and precisely adjusted, it is taken apart again. The movement parts are cleaned and decorated by hand with finishing and polishing techniques, followed by the final assembly procedure. This assures long-term functional integrity and the immaculacy of all artisanal finishes. Even if this takes a little more time. **www.alange-soehne.com**

[Summary]

178 Brand Profiles and Watch Collections

178 A. Lange & Söhne	286 IWC
184 Audemars Piguet	292 Jacob & Co.
190 Blancpain	298 Jaquet Droz
196 Breguet	302 Longines
202 Breitling	312 Louis Vuitton
206 Bulgari	320 Omega
218 Carl F. Bucherer	326 Patek Philippe
226 Cartier	334 Piaget
234 Chaumet	344 Richard Mille
238 Chopard	352 Roger Dubuis
244 de GRISOGONO	358 Rolex
252 Dior	364 TAG Heuer
258 Franck Muller	370 Vacheron Constantin
264 Girard-Perregaux	378 Zenith
270 Guy Ellia	
278 Hublot	

Ricardo Guadalupe
CEO of Hublot

FEVER PITCH

After the World Cup in 2018 and a record year for sales, in 2019, Hublot is preparing to join in celebrating the 90th anniversary of the Ferrari racing team. For Ricardo Guadalupe, CEO of Hublot, **ONE MAXIM HOLDS TRUE: NEVER REST ON YOUR LAURELS**.

What can you tell us about the business end of things in 2018?

With record-breaking sales—double-digit growth—and the various projects that we launched over the course of the year, we can safely say that 2018 was a historic year for Hublot.

How is Hublot different from other brands?

The "Art of Fusion" means we always look to the future. We possess an authentic watchmaking identity, despite our young age, and we invest considerably in research and development, both for movements and materials, while continuing to work on our vertical integration. We plan to raise our current annual production of about 15,000 Unico movements to 30,000 in the medium term. To that end, after inaugurating a second building in Nyon in 2015, we are planning a third one soon. We are increasing our facilities not necessarily in order to produce more movements, but rather to build up our expertise in the manufacture of calibers and cases. For example, we assembled 30,000 high-tech cases in materials such as ceramic, Magic Gold, and sapphire in 2018.

▶ **BIG BANG MP-11 3D CARBON**

FEVER PITCH

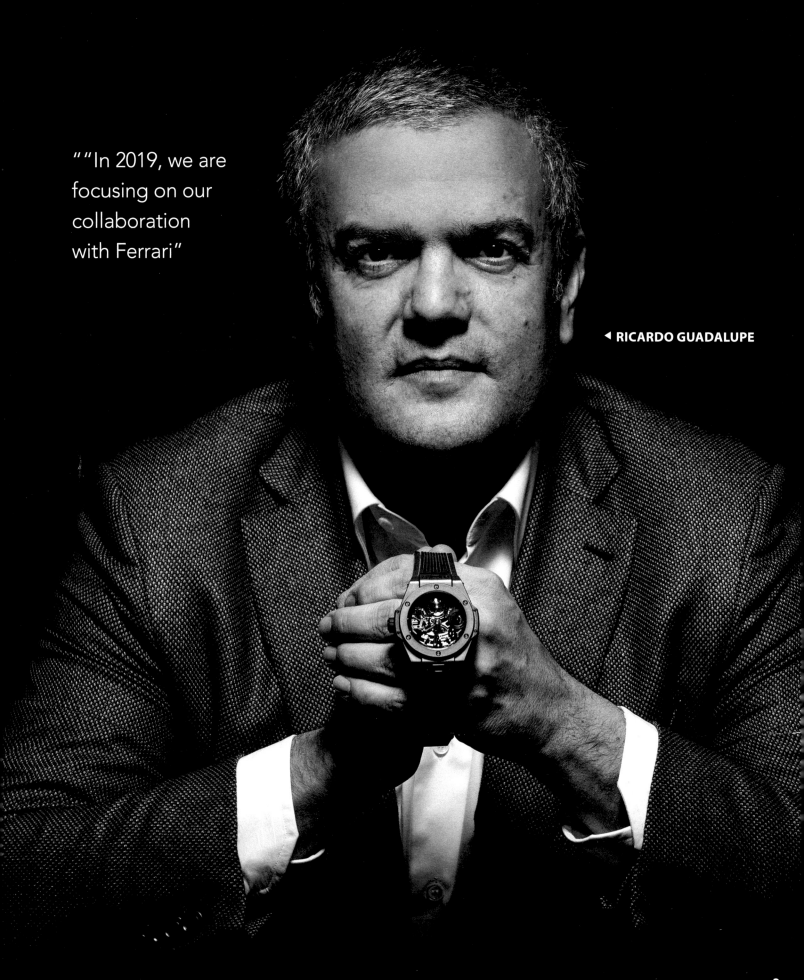

" "In 2019, we are focusing on our collaboration with Ferrari"

◀ RICARDO GUADALUPE

What were Hublot's key products in 2018?

First of all, the 2018 FIFA World Cup Russia™ Big Bang Referee limited edition was a huge success, with 2018 pieces sold in one month! We fulfilled a specific need expressed by FIFA and, for the first time, referees wore this watch with smart technology developed by Hublot. In addition to the functions available with the limited edition models, the referee's watch was connected to goal line technology. Then, with the Big Bang Unico Red Magic, we continued developing the Art of Fusion; we worked for four years on this world premiere, a bright red ceramic watch. And finally, the Big Bang MP-11, with a battery life of 14 days and a unique architecture equipped with seven series-mounted barrels redesigned to accommodate the Big Bang case, perfectly represents the innovative spirit of Hublot.

What events marked 2018?

Without a doubt, the FIFA World Cup in Russia last summer. It was a magnificent experience at the heart of the biggest sporting event in the world. In addition to more than 1,000 guests welcomed throughout the competition, we were present on the field with referee tables in our Big Bang shape. This direct exposure had an incredible impact in terms of brand recognition, which we had not even imagined when we started working with FIFA in 2010. As official timekeeper, we responded to a need by creating the Big Bang Referee 2018 FIFA World Cup RussiaTM. Finally, several coaches, including Didier Deschamps, the coach of the victorious French team, wore Hublot on their wrists. Being an integral part of matches, providing valuable information to players and viewers, gaining worldwide visibility—this is all the result of a long-term, close relationship with FIFA and the soccer community. Together, we think, develop, and improve. And nothing is more rewarding!

What were some good surprises for the brand in 2018?

The FIFA World Cup effect had a direct impact on sales! And we have partnered with international soccer star Kylian Mbappé, who succeeds Jamaican sprinter Usain Bolt. By becoming an Hublot ambassador, the PSG striker joins soccer legends such as the Brazilian Pelé, with whom he shares common values on social media. Kylian Mbappé conveys a powerful message and inspiring values: the power of dreams and passion, the magic of sport. But this partnership is above all the meeting with a beautiful person, a young man—already a legend—who embodies the future and all its possibilities.

▲ 2018 FIFA WORLD CUP RUSSIA™ REFERENCE 20

▲ BIG BANG UNICO RED MAGIC

▲ BIG BANG MP-11

◄ KYLIAN MBAPPÉ
► PELÉ

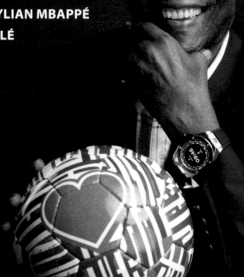

FEVER PITCH

▲ LANG LANG   ▲ SHEPARD FAIREY

What are the main themes that Hublot will develop in 2019?
2019 will focus on our collaboration with Ferrari, whom we are helping to celebrate an important milestone: the 90th anniversary of Scuderia Ferrari. The Art of Fusion is obviously a central focus, particularly in the search for new materials, such as ceramics, where we are leaders and continue to innovate, or 3D carbon for which we have discovered new applications in the watch industry. And we do not intend to stop there!

What exciting news can we expect for 2019?
Get ready for Baselworld, where we will present a revolutionary model created in collaboration with Ferrari as well as a new version of the Big Bang Sang Bleu, which takes the concept even further.

What are the strongest markets for Hublot and its distribution network?
Our great strength is that we have a balanced distribution in the main regions of the globe (Americas = 22%, Europe = 32%, Asia = 37%, Others = 9%) and so we don't depend on a country or a specific clientele. We still have huge development potential in China, where we target a younger clientele. Our selective distribution has approximately 800 points of sale, including 93 Hublot boutiques that represent one third of our sales. Since the first Hublot boutique in 2007, we have opened an average of 10 new ones per year and will maintain this pace in the years to come, while continuing to invest in improving and expanding existing stores. In particular, we renovated our flagship store in Paris's Place Vendôme and opened new ones in Geneva and London in 2018. In 2019, we are planning some openings in traditional luxury destinations, as well as in cities where we will be pioneers, such as Rome, Hong Kong or Athens, in order to maintain close relationships with our newest customers.

Tell us about your partnerships...
At Hublot, partnerships are developed based on the interests of our customers, to better meet their expectations. Soccer, cars, art and music are areas of overlapping interest for our clientele, but our two main focuses are partnerships with FIFA and UEFA, as well as with Ferrari. We also continue our collaborations in the world of art, with the street artist Shepard Fairey or the piano virtuoso Lang Lang. True to our motto "Be first, unique, different," we do things differently than others and always seek to bring our world together with that of our ambassadors.

What are the main challenges for the future of watchmaking?
Above all, never resting on one's laurels, but instead reinventing the art of watchmaking and positioning oneself as a leader in this new universe. Nobody buys a watch just to display the time anymore, so it has to become a piece of art in its own right, a statement for the wrist.

What is your position regarding online sales and social media?
Often, customers who buy a watch on the Internet are looking for discounts. Chrono24 is a good example, with discounts of 10-20%, or even more, or new ways of financing like the recent leasing offers offered on these sites. Launching an online sales site is not enough; one must offer a real customer experience using digital tools, integrated into a specialized network, with experienced resellers and professional implementation. We must recreate a service identical to that which one might expect when crossing the threshold of an Hublot shop. Social media, including Instagram, have taken on an important role in brand marketing. Thanks to them, we can be closer to our customers, share our news, and offer them a great experience with our brand.

HUBLOT

Big Bang MP-11 Power Reserve 14 Days 3D Carbon

Always on the hunt for new, creative materials to transform with a brilliantly conceived timepiece, Hublot has produced the first 3D carbon case, composed of polymer matrix with three-dimensional fibers. Ultra-lightweight—the entire watch, including the rubber strap, weighs just 90g—and highly resistant, the case provides a stunning framework for a spectacular movement. The 3D carbon was previously reserved for military use, and the unique striations of the dark material testify to its rugged, ready-for-anything pedigree. The unusually shaped case ripples outward across the bottom of the dial, shrinking the time indication to make room for the star of the MP-11: the seven series-coupled barrels and the impressive 14-day power reserve (and the display thereof) that they drive. The curvature of the sapphire crystal over this power reserve display creates a magnifying effect, emphasizing its importance. Smoked, transparent composite sides offer further perspectives on the HUB9011 movement, which is wound by hand using the large fluted crown or a racing-inspired, electric Torx stylus.

Big Bang Unico GMT Carbon

Hublot also uses carbon fiber in the case of the strikingly attractive Big Bang Unico GMT Carbon. The conception of the dial is more traditional than the MP-11, but Hublot still puts an idiosyncratic spin on the proceedings, matching the blue composite resin bezel with the structured lined rubber strap that bears the timepiece, as well as the day/night ring that corresponds to the 12-hour second time zone. The 45mm carbon-fiber case and bezel sport a checkerboard pattern that lends a masculine texture and point of visual intrigue to the piece. The rectangular pushbuttons on the side of the case at 2 and 4 o'clock advance and retreat the local time, respectively, while the home time, displayed via an arrow-shaped pointer, is set using the crown. The simplicity and ease of use for this arrangement is a huge advantage for people who have to travel often—which is to say, a great many of us! Ricardo Guadalupe summed up the approach nicely: "Every complex problem has a simple solution. It simply depends on how we look at it. I travel so much that, as often as not, my watch is the only way I can keep track of what time it is at home, and where I am landing. And, I must admit, I have never found a GMT model which is simple enough to really be useful, and that I can use as intended. Thanks to the ingenuity of our R&D team, this is now possible and, believe me, the Big Bang Unico GMT is sure to inspire imitators."

Big Bang Unico King Gold Ceramic

Always on the lookout for unusual new materials with unexpected benefits, Hublot started the trend with its 18-karat King Gold, an alloy that increased the percentage of copper and added platinum to a traditional gold alloy to stabilize the color long-term and neutralize oxidation. The result is not only a more durable watch, but a more beautiful one: its exclusive color is warmer and redder than the traditional 18K gold 5N. The Big Bang Unico King Gold Ceramic combines an 18-karat King Gold case with a black ceramic bezel for a robust take on the flyback chronograph. Hublot's signature H-shaped titanium screws punctuate the bezel, securing the ceramic layer to the anodized black aluminum lower bezel and the King Gold case. Red accents make the openworked dial easy to read at a glance. An affirmation of the timepiece's sporty personality, the black structured rubber strap provides a fuss-free method of wearing this assertive Big Bang model.

Big Bang Unico Red Magic

The creation of vibrantly colored ceramic has been something of a Holy Grail for Hublot ever since the company invented Magic Gold, an alloy of 24-karat gold and boron carbide. After four years of research, Hublot's R&D department developed a process that could use intense pressure and heat, without damaging the bold pigments needed. Patented by Hublot and exclusive to the brand, this colored ceramic is innovative in and of itself, as well as in its manufacturing process and its increased hardness vis-à-vis regular ceramics. The hardness and resilience of this material also stems from its extreme density: 1500 HVI, versus 1200 HVI with conventional ceramics.

Starting off with a bang, Hublot has chosen a bright fire engine red to kick off the introduction of this groundbreaking material to its palette. The generous 45mm case diameter makes the visual impact even more powerful. The openworked dial reveals the activity of the 330-component HUB1242 UNICO automatic chronograph, another Hublot innovation, whose dark hues contrast with the watch's bright red flange, indexes, minutes and seconds counters, Arabic numerals, and hands. Released in a limited edition of just 500 pieces, the Big Bang Unico Red Magic is a harbinger of things to come, as the brand is sure to expand its research across the rainbow.

BIG BANG

Big Bang Referee 2018 World Cup Russia

As the official timekeeper of the FIFA World Cup™, Hublot is an enthusiastic partner, constantly coming up with new ways to express the relationship through horology. With its Big Bang Referee 2018 World Cup Russia™, the Swiss brand designed the perfect watch for the referees judging the plays on this global stage, as well as soccer fans around the globe. Its 49mm diameter makes it an arresting accessory, while allowing for a larger display that is instantly readable in high-pressure situations. In a first for Hublot, this model is a connected watch, and its wearers experienced a panoply of alerts for relevant alerts, beginning with a countdown of the days until the event itself. The watch notified its wearers 15 minutes before the start of each match, as well as yellow and red cards, player changes, and goals—this last noted in real time, vibrating and displaying the word "GOAL" whenever a point was scored. The actual referees walked out on the field with their watches connected to goal-line technology that followed all the trajectories of the ball, determining beyond a doubt if the ball did or did not entirely cross the goal line. Among many other unique features, this connected watch is available with two neutral dial designs, or 32 designs inspired by the flags of the competing countries. Compatible with the Android ecosystem, there is no end to the customization options of this watch and the applications available to it, and its battery can be recharged simply by placing it on a contact charger. Available in a limited edition of 2,018 models, this piece is delivered with two straps: one in Hublot's classic black lined rubber, and a cuff strap adorned with the 2018 FIFA World Cup™ emblem. Also available are straps in the colors of each country. The brains of this piece are powered by INTEL® Wear OS by Google technology developed by LVMH.

GRACE AND CHARACTER

Joséphine Collection

CHAUMET
PARIS

L'art de la joaillerie depuis 1780

Westime

From international timepiece collectors, to customers who are considering their first purchase of a watch, Westime's four boutiques in Southern California and Miami cater to them all by offering a broad and deep selection of today's most desirable watches.

▲ Westime La Jolla ▲ Westime Beverly Hills

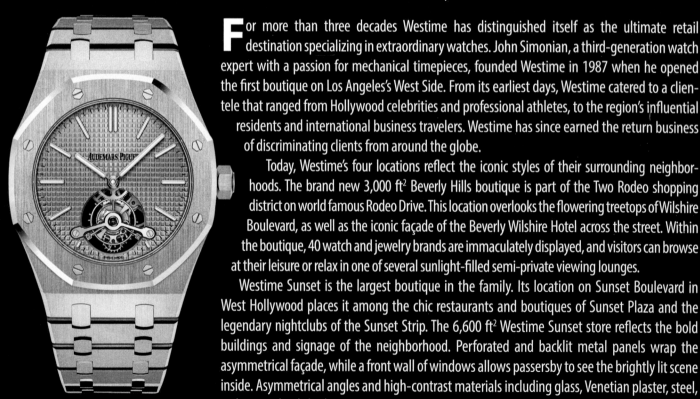

▲ **ROYAL OAK TOURBILLON EXTRA-THIN WESTIME LIMITED EDITION**
(Audemars Piguet)

For more than three decades Westime has distinguished itself as the ultimate retail destination specializing in extraordinary watches. John Simonian, a third-generation watch expert with a passion for mechanical timepieces, founded Westime in 1987 when he opened the first boutique on Los Angeles's West Side. From its earliest days, Westime catered to a clientele that ranged from Hollywood celebrities and professional athletes, to the region's influential residents and international business travelers. Westime has since earned the return business of discriminating clients from around the globe.

Today, Westime's four locations reflect the iconic styles of their surrounding neighborhoods. The brand new 3,000 ft^2 Beverly Hills boutique is part of the Two Rodeo shopping district on world famous Rodeo Drive. This location overlooks the flowering treetops of Wilshire Boulevard, as well as the iconic façade of the Beverly Wilshire Hotel across the street. Within the boutique, 40 watch and jewelry brands are immaculately displayed, and visitors can browse at their leisure or relax in one of several sunlight-filled semi-private viewing lounges.

Westime Sunset is the largest boutique in the family. Its location on Sunset Boulevard in West Hollywood places it among the chic restaurants and boutiques of Sunset Plaza and the legendary nightclubs of the Sunset Strip. The 6,600 ft^2 Westime Sunset store reflects the bold buildings and signage of the neighborhood. Perforated and backlit metal panels wrap the asymmetrical façade, while a front wall of windows allows passersby to see the brightly lit scene inside. Asymmetrical angles and high-contrast materials including glass, Venetian plaster, steel, walnut, and polished concrete create a gallery-like setting inside the two-story space. Custom corners for Audemars Piguet, Breitling, Omega, TAG Heuer, Zenith, Bulgari, and Buben & Zörweg enhance the shopping experience for fans of those popular brands.

WESTIME

▼ Westime Sunset

▲ Breitling boutique Los Angeles   ▲ Westime Miami   ▲ Westime Miami

▲ **NAVITIMER DAY & DATE**
(Breitling)

▶ **FREAK OUT**
(Ulysse Nardin)

Westime La Jolla is located north of San Diego in one of the country's most beautiful seaside communities. The light-filled store resides among charming shops, galleries and cafes on elegant Prospect Street, just steps from the Pacific Ocean. The boutique's gray slate flooring, natural wood and glass watch cases, and open floor plan invite customers to browse casually. The experienced Westime staff provides such special services as watch repairs and water resistance tests.

Westime selected South Florida for its first location outside of California, and opened Westime Miami in the city's new downtown shopping mecca, Brickell City Centre. The showroom interconnects with neighboring Audemars Piguet and Richard Mille boutiques, with a rotunda at its heart to showcase extraordinary pieces.

Westime is frequently noted as one of a dozen multi-brand retailers in the world that influences trends in the watchmaking industry. Led by John Simonian and his son Greg, the company is dedicated to offering the most important creations from traditional watch brands, while also promoting the new guard in haute horology. At Westime's boutiques, there is always something new to discover.

Westime also operates Audemars Piguet and Breitling boutiques in select cities. The company is proud to support numerous charitable causes, including After-School All-Stars.

▼ Greg Simonian, Arnold Schwarzenegger, Patrick Schwarzenegger at the Westime Charity Event to benefit After-School All-Stars, Los Angeles

▼ Westime also supported last year's charity event in Miami supporting After-School All-Stars

WESTIME

▲ AMC (Urwerk)

▲ GRANT (MB&F)

BRANDS CARRIED

AUDEMARS PIGUET	HAMILTON
BELL & ROSS	HARRY WINSTON
BLANCPAIN	HAUTLENCE
BREGUET	H. MOSER & CIE
BREITLING	HUBLOT
BRM	HYT
BUBEN & ZORWEG	LONGINES
BULGARI	LOUIS MOINET
CARL F. BUCHERER	MAITRES DU TEMPS
CHOPARD	MB&F
CHRISTOPHE CLARET	MCT
DE BETHUNE	MESSIKA
DE GRISOGONO	MONTEGRAPPA
DEVON	NIXON
DEWITT	OMEGA
DIOR	REUGE
DOTTLING	RICHARD MILLE
ERNST BENZ	ROLAND ITEN
FABERGE	RUDIS SYLVA
FIONA KRUGER	SHAMBALLA JEWELS
FRANCK MULLER	TAG HEUER
FREDERIQUE CONSTANT	TISSOT
GIRARD-PERREGAUX	TUDOR
GIULIANO MAZZUOLI	ULYSSE NARDIN
GRAND SEIKO	URWERK
GLASHUTTE ORIGINAL	ZENITH
GREUBEL FORSEY	

▼ ▶ MESSIKA JOAILLERIE

BEVERLY HILLS
206 North Rodeo Drive
Beverly Hills, CA 90210
T: 310-888-8880

MIAMI
Brickell City Centre
701 South Miami Avenue, Suite 167C
Miami, FL 33131
T: 786-347-5353

LA JOLLA
1227 Prospect Street
La Jolla, CA 92037
T: 858-459-2222

SUNSET
8569 West Sunset Boulevard
West Hollywood, CA 90069
T: 310-289-0808

info@westime.com / www.westime.com

Greg Simonian
President of Westime

GUIDING LIGHT

As the President of Westime, Greg Simonian has **SPENT A DECADE SENSING THE NEWEST TRENDS BEFORE THEY SWEEP THE INDUSTRY, AND WORKING WITH BRANDS TO DEVELOP MUTUALLY BENEFICIAL STRATEGIES**. Having grown up in the business, Simonian has seen decades' worth of changes transform the watch world from top to bottom. He shares his unique perspective on Westime's future and the industry as a whole.

Where do you spend most of your time, and how? What is your favorite part of running Westime?

As our company has expanded, and added standalone monobrand boutiques as well as our four multi-brand boutiques, I am increasingly involved in lease negotiations and business contracts! But thankfully I still have plenty of involvement in my favorite part of the business, which is discovering new watches directly from the fascinating people who created them.

Westime is strongly linked with California—it's even in your name. How do you see that relationship evolving in the years ahead?

It remains our heart and soul, and in terms of business it is where our corporate headquarters is located, and the majority of our boutiques. I grew up in California, I root for the local teams, and my family is here so we are grounded! But just as our customers are world travelers and often homeowners in various parts of the country, Westime is comfortable bringing its particular brand of service and style to additional destinations, like we did in Miami.

GUIDING LIGHT

"Westime makes it easy, comfortable, educational and never intimidating to get the lay of the land and discover a vast universe of watches."

▶ GREG SIMONIAN

GUIDING LIGHT

Westime Beverly Hills recently moved to a new location, just a few doors away from the former one. What was the impetus behind this move? How does this open new doors for Westime and your mission?

When the space became available overlooking Wilshire Blvd., we immediately saw the potential for Westime. There is more room, for starters, so each collection we carry can be displayed beautifully. The wall of windows overlooking Wilshire is quite spectacular, and brings in the most refreshing California sunshine throughout the day. And additional benefit is that Westime can be seen by guests staying and dining at the Beverly Wilshire Hotel across the street. The luxury hotel's customers have always been great clients of Westime through the years.

In addition to Westime's physical locations, the brand is placing a strong emphasis on Westime.com. What kind of experience are people looking for when they shop online? How does the website complement the brick-and-mortar stores, and vice versa?

The new Westime.com is easier to navigate, more intuitive, filled with new watch pictures and news. Followers have easy and direct access to our staff through the site, so their questions can be answered immediately. With a marketing program that supports the site, visitors searching the Internet for a particular watch can quickly see when it is available for sale at a Westime location.

Second-hand sales are becoming a bigger part of the watch market, and Westime has taken steps toward joining this trend. What advice would you give to someone who is interested in buying a previously owned timepiece?

The most important piece of advice is, buy from a trusted source. If you buy directly from an unknown seller over the Internet, you have little recourse if the watch you get turns out to be inauthentic or damaged.

There does not seem to be a consensus in the industry on whether multi-brand retailers or mono-brand boutiques are more desirable. What are your thoughts on the subject?

There is definitely a trend among popular watch brands to shift away from multi-brand points of sale, such as Westime, and open their own network of standalone boutiques. Westime is involved in both sides of the equation, since we are partners now with Audemars Piguet and Breitling to run some of their mono-brand boutiques. But there remains a strong need for the multi-brands: both from the standpoint of a watch company that does not want its own boutiques around the world as its business strategy, and also from the standpoint of customers. Someone new to the world of watches may not know where to start when they see a boulevard of individually branded watch boutiques! But Westime makes it easy, comfortable, educational and never intimidating to get the lay of the land and discover a vast universe of watches.

Westime has helped many customers find their first luxury watches, and those customers keep coming back! What is a popular first watch that someone might buy, and what are they more likely to acquire after building a collection?

There really isn't a cookie-cutter template for watch collectors, which is a beautiful thing! I could say that a more affordable watch is a popular starting point, such as a TAG Heuer, Longines, Tudor or Hamilton. But from there a collection could branch in so many directions! We have some clients who love to collect one type of complication—such as perpetual calendars—from many different manufacturers. Others fall in love with one special watch and seek to acquire it in every color and limited edition that emerges. To each his own!

How has the watch industry changed in the last ten years? The last five?

Over that period excellence in watchmaking has endured. But the business of buying and selling them is changed significantly. The internet, of course, changed retail forever, and if you include social media with it, the ways that people communicate about watch news and make their purchases is much different than it was a decade ago. That said, Westime has four brick-and-mortar stores, with some sales professionals who have been on our team for nearly 30 years. There is something to be said for longevity and expertise.

What do you predict will be the big trend to emerge in 2019?

I predict there will be increased interest in watch sales at the big auction houses around the world, especially when timepieces of historic importance come up for sale.

[Web Site Directory]

A. LANGE & SÖHNE	www.alange-soehne.com
AUDEMARS PIGUET	www.audemarspiguet.com
BLANCPAIN	www.blancpain.com
BREGUET	www.breguet.com
BREITLING	www.breitling.com
BVLGARI	www.bulgari.com
CARL F. BUCHERER	www.carl-f-bucherer.com
CARTIER	www.cartier.com
CHAUMET	www.chaumet.com
CHOPARD	www.chopard.com
DE GRISOGONO	www.degrisogono.com
DIOR	www.dior.com
FRANCK MULLER	www.franckmuller.com
GIRARD-PERREGAUX	www.girard-perregaux.com
GUY ELLIA	www.guyellia.com
HUBLOT	www.hublot.com
IWC	www.iwc.com
JACOB & CO.	www.jacobandco.com
JAQUET DROZ	www.jaquet-droz.com
LONGINES	www.longines.com
LOUIS VUITTON	www.louisvuitton.com
OMEGA	www.omegawatches.com
PATEK PHILIPPE	www.patek.com
PIAGET	www.piaget.com
RICHARD MILLE	www.richardmille.com
ROGER DUBUIS	www.rogerdubuis.com
ROLEX	www.rolex.com
TAG HEUER	www.tagheuer.com
VACHERON CONSTANTIN	www.vacheron-constantin.com
ZENITH	www.zenith-watches.com

RELATED SITES

BASELWORLD	www.baselworld.com
SIHH	www.sihh.org

AUCTION HOUSES

CHRISTIE'S	www.christies.com
SOTHEBY'S	www.sothebys.com

BREITLING BOUTIQUE
NEW YORK • MIAMI • LOS ANGELES
SCOTTSDALE • DENVER • HOUSTON
WASHINGTON DC • SAN ANTONIO
ORLANDO • LAS VEGAS

Georges Kern
CEO of Breitling

OPEN TO ALL

Under the leadership of Georges Kern, CEO since 2017, **BREITLING HAS BEEN MAKING A PLACE FOR ITSELF IN THE LUXURY SPORTS WATCH SECTOR**. Observers estimate its growth at between 10 and 15% in 2018. The brand now conveys its "modern retro" philosophy (in the words of its CEO) both in its communication and in the design of models. One example of this is its capsule collection, launched at the beginning of the year and dedicated to civil aviation from the 1950s to the '70s. Another difference between Breitling and other high-end brands is that Breitling's watchmaking luxury is "inclusive," rather than "exclusive." The proof is in the very architecture of its boutiques, an "industrial loft" style that invites all to enter and engage. Finally, the brand is making its mark on the Chinese market and the digital space—with its "huge potential for growth!" Kern emphasizes.

What were Breitling's key achievements in 2018?

First, we improved the cohesiveness of our collections, thanks to Navitimer 8 and Premier. In distribution, we began integrating Breitling agents around the world. This huge verticalization project will improve our margins and make us more responsive, as we get closer to our customers. In terms of markets, the brand is making its entry into China, particularly with the conclusion of agreements with Tmall.com (Alibaba Group). Our online presence was previously almost non-existent. Moreover, in the area of communication and marketing, the new concept store is proving a great success. We also presented our "Squads," which bring together exceptional individuals from disparate fields around a central theme.

What developments has the business undergone?

The Skyracer, a quartz entry-level model priced at 2000 CHF, is no longer part of our collections; the average price of our models has increased. In general, we have reduced the number of different models we offer, better segmented our products, and created less bulky, more classic/timeless watches adapted to all wrists, including the smallest watches for both women and men. The preliminary results are really encouraging.

▶ **PREMIER B01 CHRONOGRAPH 42 BENTLEY BRITISH RACING GREEN**

OPEN TO A

"The Breitling universe is inclusive, and its production is exclusive."

◀ GEORGES KERN

OPEN TO ALL

▲ From left to right: Peter Lindbergh, Georges Kern, Brad Pitt, and Daniel Wu

▲ Breitling Gala Night in Beijing

Which markets stand out?

Breitling is very popular in the United States, but also in England and Japan. China represents a tremendous growth opportunity for the brand, especially considering the fact that Greater China currently represents 50% of the Swiss watch industry market! At the moment, we have fewer than 25 points of sale in China. I plan to reach about 100. Around the world, we have around 80 stores, of which 20 are in-house and 60 are outside points of sale. We will continue this distribution strategy.

Has your partnership with T-mall already borne fruit?

Yes, it has enabled us to reach millions of people and will result in significant sales. Our brand ambassadors Daniel Wu and Brad Pitt also generated enormous visibility in China when we launched last November. Moreover, digital technology makes it possible to communicate directly with the consumer, with possibilities of personalization that fit perfectly into our "inclusive" approach. Online consumers are not the same people as those who prefer buying in boutiques. In addition, Chinese Millennials no longer want to buy the same watches as their parents. Familiar with large, imposing models, they prefer sports watches. This change is a new stage in the maturity of the market and opens opportunities for sport-chic watches, like the Breitling models.

How has the public responded to the Premier?

This watch is a phenomenal success. It immediately became one of our top four sellers. Our retail outlets have not even been fully supplied yet: deliveries began in mid-November. So we are expecting excellent results as it rolls out further.

How would you describe the Premier?

The Premier follows the style of the house, with several indications on the dial. Its sleek, "modern retro" design combines functionality and style, perfectly complementing our other current models, which are very sporty. The Premier enables us to reach a new clientele in Western countries, but of course also in Asia.

Could you tell us more about the "modern retro" concept?

In response to the extreme digitalization of today's world, people need a sense of roots, and they are looking for authenticity. We built the whole brand around this "modern retro" spirit. The new design of the boutiques expresses the concept perfectly. In addition, this concept is "inclusive": people must take pleasure in visiting our stores and coming inside to shop. It enables us to reach young and mature alike. Of course, our production, as it is quite limited, remains necessarily exclusive.

▶ **PREMIER B01 CHRONOGRAPH 42**

OPEN TO ALL

▲ Breitling Triathlon Squad and Friends

▲ Bertrand Piccard (*back*) and Georges Kern (*far right*) at Breitling Triathlon

Is this concept part of your communication?

Absolutely. Our partnership with Kelly Slater, for example, is part of this inclusive trend. We choose to associate with a number of extremely popular sports, such as triathlon—and thus running, cycling and swimming—along with surfing. Our ambassadors are athletes as well as accessible, credible and relatable personalities. If you listen to Bertrand Piccard's message, he talks about the environment, an important theme for today's youth. My generation spoiled the planet in recent decades, but Millennials care about it. With Kelly Slater and his eco-responsible brand Outerknown, we launched a watch featuring an Econyl strap made with fishing nets lost in the oceans. There are hundreds and hundreds of tons of this material out there! We want to raise public awareness of this problem of marine pollution. For my own part, I have been working on ecological causes for 25 years; they are very close to my heart.

In January you launched a "capsule" collection. What is that about?

This concept is inspired by the world of fashion. This is a first in watchmaking. The capsule collections can be collaborations that are limited in terms of time and production, without being numbered. They allow us to deal with very interesting themes, sometimes more fantastical or whimsical ones, but without overloading the collection and with larger volumes than in limited editions. That way we protect the real limited editions, intended for collectors. This year we are launching three pieces that reflect the golden age of civil aviation: the 1950s, '60s and '70s. The capsule collection is a tribute to legendary airlines, with the Navitimer 1 B01 Chronograph 43 Swissair Edition celebrating the former national airline of Breitling's home country, the Navitimer 1 B01 Chronograph 43 Pan Am Edition (which takes on the signature blue of the Pan American World Airways), and the beautiful TWA edition, inspired by the airline that was founded by Howard Hughes.

Finally, what do you think of the evolution of the watch industry?

The global economic and political context worries me. The stock market suffered another downturn at the end of 2018. The US and China are in a trade war; England has to contend with Brexit. Many countries are heavily indebted, and France has suffered a serious crisis with the "yellow vest" movement. In this uncertain context, brands have been able to adapt, annihilating both geographic and generational boundaries by investing in digital and addressing Millennials. Mechanical watchmaking has found its place within a market on which it now also contends with smartwatches.

▲ Kelly Slater with SuperOcean Heritage II Chronograph 44 Outerknown

▼ **SUPEROCEAN HERITAGE II CHRONOGRAPH 44 OUTERKNOWN**

▶ **NAVITIMER 1 AIRLINE EDITIONS PAN AM**

Ricardo Guadalupe
CEO of Hublot

FEVER PITCH

After the World Cup in 2018 and a record year for sales, in 2019, Hublot is preparing to join in celebrating the 90th anniversary of the Ferrari racing team. For Ricardo Guadalupe, CEO of Hublot, **ONE MAXIM HOLDS TRUE: NEVER REST ON YOUR LAURELS**.

What can you tell us about the business end of things in 2018?

With record-breaking sales—double-digit growth—and the various projects that we launched over the course of the year, we can safely say that 2018 was a historic year for Hublot.

How is Hublot different from other brands?

The "Art of Fusion" means we always look to the future. We possess an authentic watchmaking identity, despite our young age, and we invest considerably in research and development, both for movements and materials, while continuing to work on our vertical integration. We plan to raise our current annual production of about 15,000 Unico movements to 30,000 in the medium term. To that end, after inaugurating a second building in Nyon in 2015, we are planning a third one soon. We are increasing our facilities not necessarily in order to produce more movements, but rather to build up our expertise in the manufacture of calibers and cases. For example, we assembled 30,000 high-tech cases in materials such as ceramic, Magic Gold, and sapphire in 2018.

▶ **BIG BANG MP-11 3D CARBON**

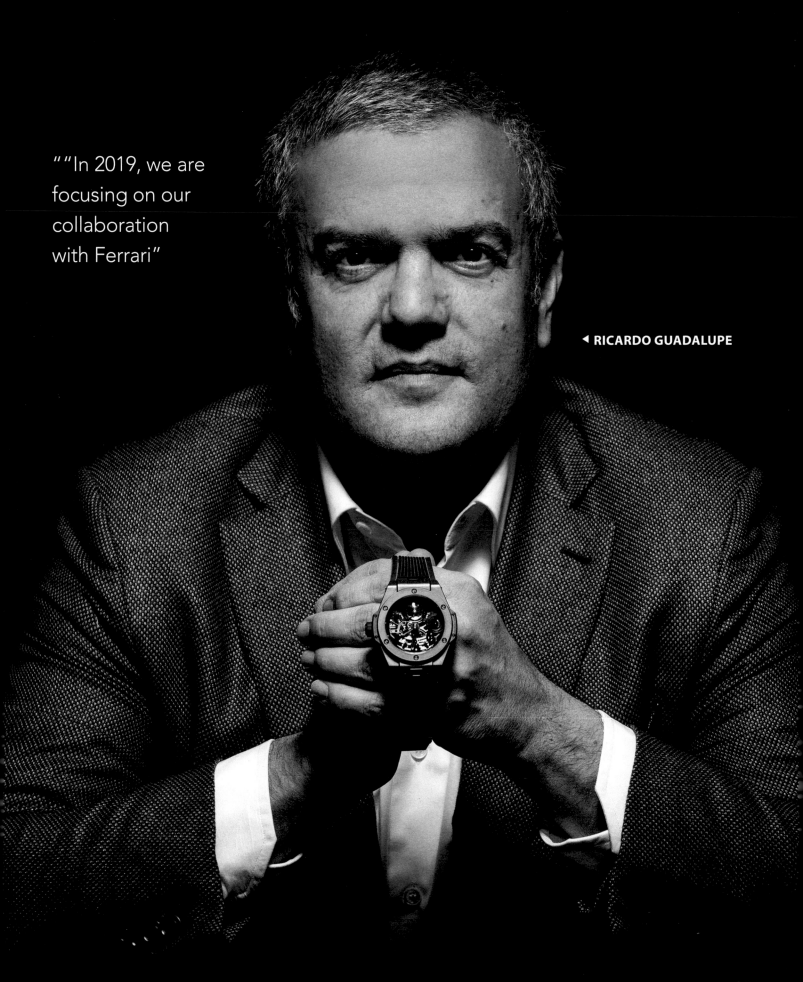

""In 2019, we are focusing on our collaboration with Ferrari"

◀ **RICARDO GUADALUPE**

▲ 2018 FIFA WORLD CUP RUSSIA™ REFERENCE 20

▲ BIG BANG UNICO RED MAGIC

▲ BIG BANG MP-11

What were Hublot's key products in 2018?

First of all, the 2018 FIFA World Cup Russia™ Big Bang Referee limited edition was a huge success, with 2018 pieces sold in one month! We fulfilled a specific need expressed by FIFA and, for the first time, referees wore this watch with smart technology developed by Hublot. In addition to the functions available with the limited edition models, the referee's watch was connected to goal line technology. Then, with the Big Bang Unico Red Magic, we continued developing the Art of Fusion; we worked for four years on this world premiere, a bright red ceramic watch. And finally, the Big Bang MP-11, with a battery life of 14 days and a unique architecture equipped with seven series-mounted barrels redesigned to accommodate the Big Bang case, perfectly represents the innovative spirit of Hublot.

What events marked 2018?

Without a doubt, the FIFA World Cup in Russia last summer. It was a magnificent experience at the heart of the biggest sporting event in the world. In addition to more than 1,000 guests welcomed throughout the competition, we were present on the field with referee tables in our Big Bang shape. This direct exposure had an incredible impact in terms of brand recognition, which we had not even imagined when we started working with FIFA in 2010. As official timekeeper, we responded to a need by creating the Big Bang Referee 2018 FIFA World Cup RussiaTM. Finally, several coaches, including Didier Deschamps, the coach of the victorious French team, wore Hublot on their wrists. Being an integral part of matches, providing valuable information to players and viewers, gaining worldwide visibility—this is all the result of a long-term, close relationship with FIFA and the soccer community. Together, we think, develop, and improve. And nothing is more rewarding!

What were some good surprises for the brand in 2018?

The FIFA World Cup effect had a direct impact on sales! And we have partnered with international soccer star Kylian Mbappé, who succeeds Jamaican sprinter Usain Bolt. By becoming an Hublot ambassador, the PSG striker joins soccer legends such as the Brazilian Pelé, with whom he shares common values on social media. Kylian Mbappé conveys a powerful message and inspiring values: the power of dreams and passion, the magic of sport. But this partnership is above all the meeting with a beautiful person, a young man—already a legend—who embodies the future and all its possibilities.

◀ KYLIAN MBAPPÉ
▶ PELÉ

FEVER PITCH

▲ LANG LANG   ▲ SHEPARD FAIREY

What are the main themes that Hublot will develop in 2019?

2019 will focus on our collaboration with Ferrari, whom we are helping to celebrate an important milestone: the 90th anniversary of Scuderia Ferrari. The Art of Fusion is obviously a central focus, particularly in the search for new materials, such as ceramics, where we are leaders and continue to innovate, or 3D carbon for which we have discovered new applications in the watch industry. And we do not intend to stop there!

What exciting news can we expect for 2019?

Get ready for Baselworld, where we will present a revolutionary model created in collaboration with Ferrari as well as a new version of the Big Bang Sang Bleu, which takes the concept even further.

What are the strongest markets for Hublot and its distribution network?

Our great strength is that we have a balanced distribution in the main regions of the globe (Americas = 22%, Europe = 32%, Asia = 37%, Others = 9%) and so we don't depend on a country or a specific clientele. We still have huge development potential in China, where we target a younger clientele. Our selective distribution has approximately 800 points of sale, including 93 Hublot boutiques that represent one third of our sales. Since the first Hublot boutique in 2007, we have opened an average of 10 new ones per year and will maintain this pace in the years to come, while continuing to invest in improving and expanding existing stores. In particular, we renovated our flagship store in Paris's Place Vendôme and opened new ones in Geneva and London in 2018. In 2019, we are planning some openings in traditional luxury destinations, as well as in cities where we will be pioneers, such as Rome, Hong Kong or Athens, in order to maintain close relationships with our newest customers.

Tell us about your partnerships…

At Hublot, partnerships are developed based on the interests of our customers, to better meet their expectations. Soccer, cars, art and music are areas of overlapping interest for our clientele, but our two main focuses are partnerships with FIFA and UEFA, as well as with Ferrari. We also continue our collaborations in the world of art, with the street artist Shepard Fairey or the piano virtuoso Lang Lang. True to our motto "Be first, unique, different," we do things differently than others and always seek to bring our world together with that of our ambassadors.

What are the main challenges for the future of watchmaking?

Above all, never resting on one's laurels, but instead reinventing the art of watchmaking and positioning oneself as a leader in this new universe. Nobody buys a watch just to display the time anymore, so it has to become a piece of art in its own right, a statement for the wrist.

What is your position regarding online sales and social media?

Often, customers who buy a watch on the Internet are looking for discounts. Chrono24 is a good example, with discounts of 10-20%, or even more, or new ways of financing like the recent leasing offers offered on these sites. Launching an online sales site is not enough; one must offer a real customer experience using digital tools, integrated into a specialized network, with experienced resellers and professional implementation. We must recreate a service identical to that which one might expect when crossing the threshold of an Hublot shop. Social media, including Instagram, have taken on an important role in brand marketing. Thanks to them, we can be closer to our customers, share our news, and offer them a great experience with our brand.

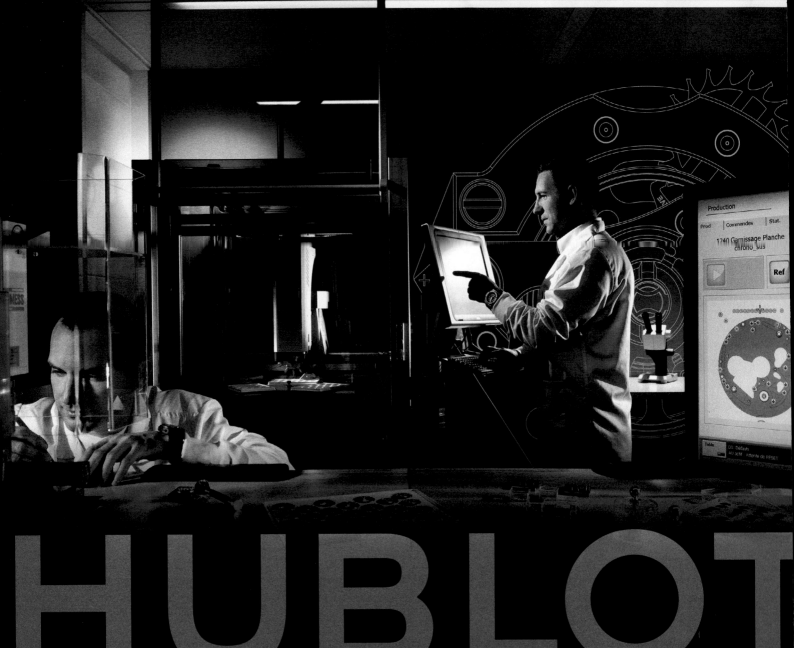

Big Bang MP-11 Sapphire.
Scratch-resistant sapphire case. In-house manual-winding
power reserve movement with 7 series-coupled barrels and a
14-day power reserve. Limited edition of 200 pieces.

LONGINES

Elegance is an attitude

Kate Winslet

Walter von Känel
CEO of Longines

STEADY WORLDWIDE GROWTH

The Longines brand has been registered in Switzerland since 1889 and around the world since 1893. **IT HAS BECOME THE WORLD'S OLDEST TRADEMARK STILL IN ACTIVE USE, WITHOUT BEING MODIFIED, IN THE REGISTERS OF THE WORLD INTELLECTUAL PROPERTY ORGANIZATION.** This classes it among the historical players in Swiss watchmaking. The brand is clearly the leader in its price segment, namely timepieces retailing for between 1500 and 3000 Swiss francs. In 2018, one of the most productive years in its history, Longines continued to grow. The Swatch Group brand owes this success to a solid strategy, overseen by the man at the helm for 30 years: Walter von Känel. Sales, production, sports partnerships, and even tracking real-time gray market sales and counterfeit products: the experienced CEO keeps a close eye on everything.

What major challenges did you face in 2018?

First of all, we have maintained the excellent dynamic that we've had for several years now. But it is important to note also that we have engaged in a relentless struggle against the gray market and fake watches. These watches, sold outside of our official system, have circumvented customs taxes and VAT, and fuel criminal networks. This system weakens all actors. By purchasing ourselves Longines watches sold on the internet—to the tune of 30 to 40 per month—we were able to track down the retailer to whom we originally sold the piece. If it wasn't sold where it should have been, we take the appropriate measures. When it comes to counterfeit goods, we send the information to the proper authorities. We also always check the authenticity of the warranty card, which is also subject to counterfeiting. Taking action against the gray market naturally has a collateral effect on sales, and in some markets, we have felt a decline. But it's a necessary evil, and we will continue to fight as fiercely as ever.

How was 2018 in terms of sales?

We achieved a turnover higher than that of 2017. It could be one of our best years, with some record months. Our production volume consistently exceeds 1.5 million euros—and recently by quite a bit—while our average sales price remains stable.

◀ **CONQUEST V.H.P. GMT**

STEADY WORLDWIDE GROWTH

" Longines is engaged in a running battle with parallel markets and counterfeiting"

▶ **WALTER VON KÄNEL**

STEADY WORLDWIDE GROWTH

◀ RECORD

Which models have particularly distinguished themselves?

The public reacted very positively to the Conquest V.H.P., launched in February 2017. This is the best quartz caliber we have ever made—V.H.P. means Very High Precision. Its outstanding accuracy varies by no more than ± 5 seconds per year. Its perpetual calendar avoids any need for date correction until 2399. It also has a magnetic field detection system with stop compensation and a "Gear Position Detection" system to control shocks. All this for a price of less than 1,000 CHF. To our great satisfaction, the Conquest V.H.P. is also quite popular in China, which is not usually a market for men's quartz watches. For the moment, the strongest demand is for the three-hand version and its perpetual calendar, but we also started delivering GMT and chronograph models around the end of 2018.

This is a rather technical watch, probably making it hard to explain and therefore difficult to sell?

This watch is certainly complex in terms of the technology involved, but it is very simple to use, although you still have to know the right way to do it! A brief instruction manual accompanying the watch makes it particularly user-friendly. We are also constantly training salespeople so that they can provide customers with the right information.

What other models were particularly popular?

Young people very much appreciate the HydroConquest, dedicated to water sports enthusiasts. The favorite is the new version with a mechanical movement and ceramic bezel. But this very complete line also includes versions with aluminum bezels, as well as quartz models.
We have also achieved excellent results with the men's and women's models in the Record collection, which are COSC-certified and feature a caliber equipped with a silicon balance spring. Finally, The Longines Master Collection continues to be one of our flagships.

STEADY WORLDWIDE GROWTH

Which product will take center stage in 2019?
The Conquest V.H.P. will continue to be our main focus. The demand for this exceptional caliber is considerable. We will consolidate sales and increase deliveries.

Longines remains strongly present in the sports world, with the exception of the French Open, with which you recently severed ties…
Yes, we declined the new partnership terms. This beautiful experience lasted 11 years. We were able to establish tournaments for young people, with national selections, which did not exist before. Each partner is moving forward with a positive spirit. On the other hand, we continue our long-standing collaboration with André Agassi and Stefanie Graff and their respective charitable foundations for children. And we are still closely linked to the world of horses, downhill skiing, gymnastics, archery, as well as the Commonwealth Games.

How is your distribution network set up?
Longines remains widely distributed around the world through nearly 4,000 points of sale, including both multi-brand retailers and Longines boutiques, whether owned by us or operated under franchise. These Longines single-brand boutiques generally offer a wider range than a multi-brand retailer can offer, and we of course ensure the quality of the staff. But all these formulas are complementary.

How are the different markets progressing?
We are very satisfied with our current presence in the Chinese market, where we operated well before the arrival of many brands. On-site retailers are clamoring for more watches, which is a great sign! But, more generally, the global market is much more reactive—and therefore less predictable—than 10 or 20 years ago. This is an important parameter that is essential to take note of in our business. The impact of certain political decisions, of social movements, or even of acts of terrorism, is likely to deeply disturb our industry without any forewarning. In spite of my 50 years of experience—I joined Longines in 1969—it's really a new factor for me. Tourism in France has suffered from the protests that began at the end of 2018; Russia has experienced an obvious impact from international sanctions, and even the death of the King of Thailand has had a significant impact on the watch market. Such examples are everywhere, which means we must try to have balanced sales across several regions in order to continue growing steadily, as we have been doing for years at Longines.

Elegance is an attitude

Aishwarya Rai

Record collection

Richard Mille
CEO of Richard Mille

RUNAWAY POPULARITY

Last year, the Richard Mille brand saw a sharp rise in sales revenue: 300 million CHF, corresponding to 4,600 watches produced. **THE FOUNDER'S CONSISTENT APPROACH ALLOWS HIM TO LOOK TO THE FUTURE WITH SERENITY.** "Demand is exploding and extremely powerful, but I deliberately decided, right from the start, that we would not attempt to meet it. We remain true to this principle by only partially meeting demand."

In just a few words, how would you describe the year 2018 for Richard Mille?
Overall, we had a good year, which went according to plan. In 2017, we achieved a turnover of 260 million CHF, and it exceeded 300 million in 2018. However, our main problem remains: we do not have enough watches to meet the ever-increasing demand. In 2017, we produced around 4,000 watches and some 4,600 last year. However, we are unable to fill our store windows because, as soon as they are produced, the watches are immediately acquired by our customers. Although we are increasing production by 15% a year, demand is growing much faster. With an average retail price of more than 200,000 CHF, we are in a market segment whose limits are difficult to determine and no study offers precise statistics in this respect.

◀ RM 70-01

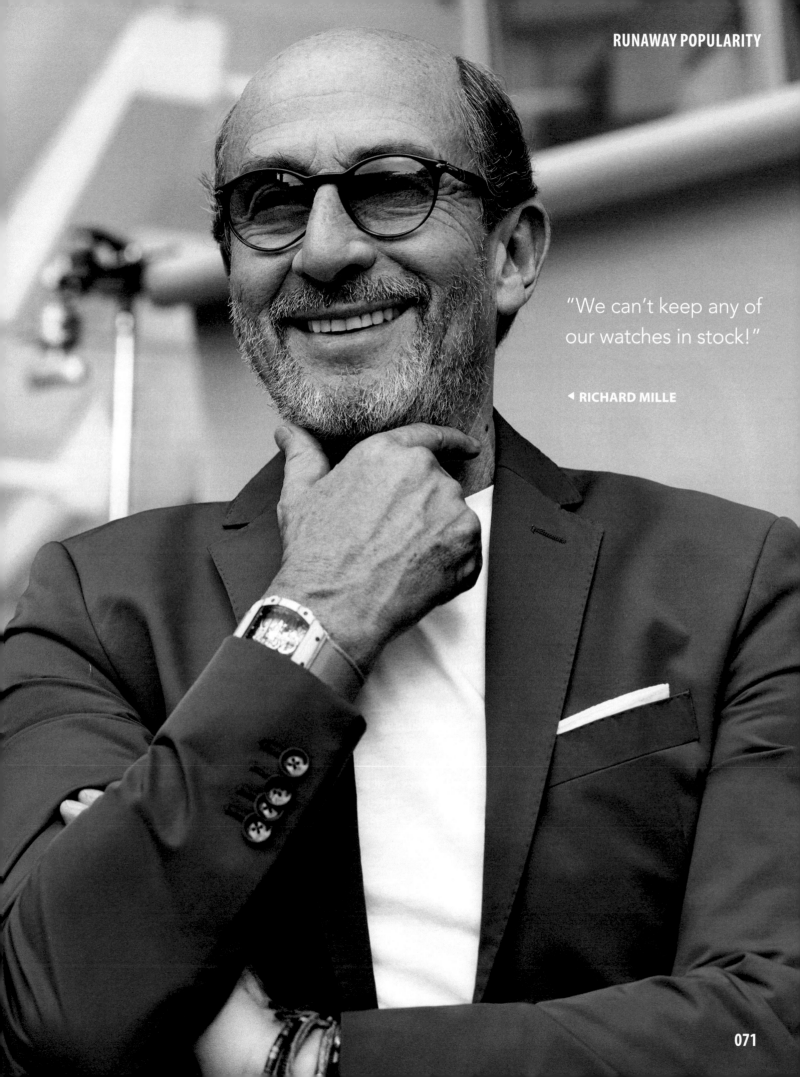

RUNAWAY POPULARITY

"We can't keep any of our watches in stock!"

◀ RICHARD MILLE

RUNAWAY POPULARITY

What were the highlights of 2018?

The year was marked by the acceleration of the development of women's watches, like the RM 71-01 Automatic Tourbillon Talisman, designed by our director of women's collections Cécile Guenat. These high jewelry, high-end watchmaking creations combine Art Deco inspirations and tribal art. The women's segment currently represents more than 25% of our sales for a production of about 900 timepieces; but it continues to grow. It is in fact proving just as impossible to keep in stock as our other collections! Our goal is to build on all ranges, yet demand is strong for all our models, whether chronographs, ultra-thin, or emblematic models such as "Baby Nadal", making it hard to satisfy everyone within such a context! Another milestone of 2018 was the opening of our largest boutique in the world in New York. It was a goal since we started, but I wanted everything to be perfect. This magnificent, record-breaking shop is exactly in the right location and in the perfect space. And finally, on the subject of the flagship products of the year, I must mention the watch we made in collaboration with the polo champion Pablo Mac Donough, equipped with an ultra-resistant laminated sapphire crystal and a cable-suspended movement.

How do you explain the difficulty of producing enough watches to meet the demand?

First of all, it's partially because they are complex watches to develop, to perfect and to manufacture. We have the goal of constantly maintaining between 800 and 1,000 watches in the pipeline, i.e. engaged in the process of manufacturing and quality control. As we operate on a tight schedule, all models that pass final quality control in December are immediately delivered at the beginning of January. We never have any watches in stock! By 2019, we plan to produce around 5,200 watches. This linear and controlled growth enables us to better organize production and regulate supply. Even though we don't meet demand, this pre-established program and controlled delivery allowed us to weather the 2008 crisis without blinking. Since our inception, demand has exploded and continues in all strategic markets, including Japan, Southeast Asia, Europe, the United States and the Middle East. Places like China and Russia promise a reservoir of untapped demand for Richard Mille.

▲ RM 71-01
▼ RM 53-01

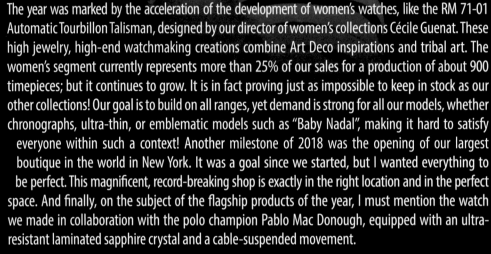

RUNAWAY POPULARITY

◄ RM 11-03

How do you explain this consistent success?

Richard Mille is a popular brand because it is not overexposed and there is a real consistency in its strategy. We could have flooded the market with cheap movements, but we want to maintain great creativity, limited production, and selective distribution. This constancy and structure are sources of reassurance and comfort for our customers. Although young, our brand exudes a sense of legitimacy and reliability, in particular with our 5-year warranty, which includes protection against shocks. Another factor contributing to our credibility is our partnerships with athletes loyal to the brand. In addition, our business model, which favors limited series, has the effect that when fans of the brand meet, none of them is wearing exactly the same model! This makes them all the more proud and happy! A Richard Mille watch is an original luxury product that is not overexposed. Finally, in recent years, our customer base is becoming younger overall, which is very encouraging for the future.

In terms of distribution and online sales, what do you have on the horizon?

Logically, we have slowed the pace of opening new stores for the reasons I mentioned above. Currently, we have around 40 shops and 20 multi-brand retailers. We will phase out the latter by the end of 2019. In the medium term, we are still planning to open stores in Beijing, Boston, Washington, as well as Dallas and Houston, and Kobe. These projects will be carried out in harmony with our pace of growth and production. As far as online sales are concerned, this channel is interesting for high-volume brands, but not for us. To date, the only Richard Mille watches sold on the Internet are usually counterfeits!

What are the key products planned for 2019?

We have many new models to unveil throughout the year, including new women's watches and a new "Bonbon" collection with neon colors.

Richard Mille no longer exhibits at the SIHH. Why does participating in the show not make sense to you now?

This event is no longer essential for the brand, because we do not do any business at the show, and in 2019, our distribution will be carried out exclusively in our own-brand stores. The orders for 2019 were completed well before the show, so our presence is no longer necessary. As far as the press is concerned, instead of large group presentations offered by the brands at a steady pace, we prefer another approach. We have recently acquired a vineyard and a country house, in the countryside of Saint-Tropez, which will be fully restored at the end of 2019. This is where we will invite our friends and journalists to share a moment of relaxation and offer them hospitality worthy of a luxury brand. We want to individualize and personalize the relationship with journalists and partners, in a warm climate.

Tell us about your collaboration with the Frieze art and culture fairs…

This partnership is a continuation of what I have always advocated: haute horology must open up to the worlds of sports, lifestyle, and art. In this vein, we already collaborate with artists to develop watches and, with Frieze, this natural bond is reinforced even more strongly. For my part, I believe that haute horology is also a form of artwork. In addition, we do not want to be classified merely as sports watches. Our presence in other universes is perfectly legitimate, because our mode of expression is not limited to a particular field.

► RM 68-01

DEFY EL PRIMERO 21

ZENITH, THE FUTURE OF SWISS WATCHMAKING

DEFYING THE ODDS

" Zenith respects the values of the past by continuing to innovate."

▶ JULIEN TORNARE

Julien Tornare
CEO of Zenith

DEFYING THE ODDS

At the head of Zenith since the summer of 2017, Julien Tornare more than believes in the iconic star of the manufacture, symbol of its history. **HIS STRATEGY IS STEERING THE BRAND TO NEW, CLEARER SKIES.** Notably, in 2018, the young CEO achieved a "sell-in"— sales to retailers—equivalent to the "sell-out"—sales to the end customer. "This is one of the commitments I gave when I arrived at the LVMH group," to which the brand belongs, Tornare says. In addition, in 2019, "Zenith will be back in the black." Coming up on the calendar of watchmaking festivities, the legendary El Primero chronograph movement celebrates its 50th anniversary. Finally, Zenith will deliver the first semi-industrial versions of the Defy Lab, a concept first presented in 2017.

Could you describe Zenith?

We bring freshness to watchmaking by drawing on our 154-year history. If all we did was make the same watches as we did over the last century, it would be easy! The Defy Lab shows how our Manufacture expresses itself in a contemporary way. To respect the values of the past, we must continue to move forward. Finally, Zenith is a real Manufacture: all our watches are equipped with our own mechanical movements.

What about Zenith by the numbers?

Zenith has nearly 300 employees. Last year, we produced just under 20,000 watches, achieving double-digit growth. This is encouraging, a sign that we are on the right track, but of course there is still a lot of work to do and we are not resting on our laurels. Still, these are the best results in a long time. The brand is returning to a healthy financial situation.

DEFYING THE ODDS

▲ EL PRIMERO 21

▼ DEFY CLASSIC
RANGE ROVER
SPECIAL EDITION 2018

What are the challenges facing Zenith this year?

This is a great year, with El Primero's 50th anniversary. A new, exceptional anniversary boxed set highlights the link between our heritage, the present and even the future. It was very successful during its presentation at the beginning of the year. The year 2019 also marks the evolution of the original El Primero caliber with the Chronomaster integrated within this boxed set. We will celebrate this fiftieth anniversary each month in a different city around the world.

Another red-letter day is the commercial launch of the Defy Lab at Baselworld. Markets are looking forward to it—and we have a lot of pre-orders. We will be able to deliver between 300 and 450 of these timepieces this year. This is a big jump from the 10 initial Defy Lab concept watches! This version is particularly successful, with the oscillator clearly visible.

Finally, our third major challenge, we are remodeling and relaunching the entire brand platform in the spring. Who is Zenith? What does its logo mean? I am convinced that strong brands rely on a campaign that generates emotion and brings the customer a sense of kinship.

Are you continuing to restrict the number of models made?

Yes, we only keep references that sell well by continuously monitoring the results. References will decrease from 178 to about a hundred this year. Thanks to this strategy, the sell-in and sell-out are perfectly aligned. Retailers aren't left with unsold stock, and we improve our turnover rate.

What kind of distribution strategy are you employing?

Zenith had 868 points of sale when I arrived. We analyzed them one by one, and 500 are enough. By the end of 2018, we had already gone down to 620. However, not all markets are equal. In the US, we still rely too little on high-quality sales points: we need to focus on attracting the elite. In contrast, in Europe, the brand has historically been present in too many retailers.

We own ten stores and ten franchised stores. To build the brand, it must be present in key cities. Apart from that, the advice available at a multi-brand retailer remains valuable. We sell hyper-technical objects, which are sometimes difficult to understand. The expertise of a sales assistant at this kind of retailer can provide objective information to the consumer. Finally, brands who are moving away from multi-brand retailers leave more market share for us. I love it!

◄ DEFY CLASSIC BLACK CERAMIC

▲ Zenith DEFY rocks centerstage at London Celebration hosted by ZENITH AND Swizz Beatz

DEFYING THE ODDS

▲ X-BAMFORD EL PRIMERO 'SOLAR-BLUE' LIMITED EDITION EXCLUSIVE TO MR PORTER

Where are you in online sales?

We sell online through partners such as Mr Porter, farfetch.com, JD.com and T-mall. Recently, we offered a Zenith X Bamford series on Mr Porter. It sold out even before it was open to the public! This success confirms my conviction that the web lends itself to special series. In December, we tested online sales on our own site, in a limited edition with Swizz Beatz. Apart from these important projects, which create buzz for our brand, beyond a certain price, online sales are not common. Especially since we do not discount prices on the Internet. This sends the customer back to the retailer.

What are your most important markets?

We achieve nearly 50% of our sales in Asia, Japan included. Europe accounts for 30%, the US for 10-12%, and the Middle East and other countries for the rest. Zenith will seek to continue its growth in Japan, and then in Greater China, as well as the USA. This last market requires substantial investment. We need to create a watch culture, including educating wealthy people that a watch displays other things (such as social status) in addition to the time! In terms of culture, Asia is already there. Customers save hard to acquire the watch of their dreams.

What are the future challenges for watchmaking?

How to preserve its appeal for new generations. Watchmaking must be careful not to be relegated to the status of old-fashioned, legacy industry. We have to make watches cool. We organize round tables with Millennials to understand their expectations. For example, young people do not see the value of acquiring a two- or three-hand watch, like that of their father or grandfather. To make them want to wear a watch—other than a smartwatch—we have to tell the exceptional stories that are part of the brand and support the current models. This approach is part of our reflection on the brand platform mentioned earlier.

I also see an issue with the distribution. In several countries, like the USA or China, the actors are the same as fifteen years ago. What will the next generation be? In addition, not all cities have quality distributors. Most brands cannot afford to open as many stores, and online sales are still limited. I'm in discussion with other brands to investigate different strategic models, such as brand associations or retailer support.

◀ PILOT TYPE 20 CHRONOGRAPH COHIBA EDITION

DEFY CLASSIC

ZENITH, THE FUTURE OF SWISS WATCHMAKING

PRECISELY

by Michel Jeannot

PRECISELY SPEAKING

SPEAKING

Although the measurement of time began as an expression of natural biological rhythms, the quest for precision has become an overriding force since the advent of mechanical watchmaking in the 14th century. In the modern world, where GPS is linked to atomic clocks and sports records are broken in increments of thousandths of a second, the idea of perfect precision is almost an obsession. Even with these technological advances, the mechanical watch still has much to say on the subject.

PRECISELY SPEAKING

Making arrangements to the nearest quarter-hour these days would be met with incredulity and even suspicion, barring the face-saving excuse of extreme traffic congestion. We live in a world where train schedules are set to the minute—at least in Switzerland!—and global positioning systems are based on information provided by atomic clocks. On the subject of the latter, in use for 50 years now, let's take a moment to remember that the latest technological advances on thermal cesium jet clocks have become so precise that they err by no more than one second every 6 million years! They are now a necessary tool for scientists, especially for the synchronization of telecommunications networks and, consequently, the "Internet of things," which will likely see 20 billion devices connected by 2020. Under these conditions, every millionth of a second counts!

◀ **REF. 5131/1P-001** (*facing page*)
(Patek Philippe)

▼ Television tower and world clock at Alexanderplatz in Berlin
(©Elena Fahro : Shutterstock.com)

astronomical knowledge, as evidenced by the great Aristotle's mention of the existence of gear mechanisms, such as the Antikythera "machine" discovered at the beginning of the last century, which recently served as a source of inspiration at Hublot. Thus, as early as the 4th century BCE, scientists could build machines displaying information with pointers on a dial. However, this intellectual abundance of antiquity, which in particular produced "our" Julian calendar in 45 BCE (developed by Sosigenes of Alexandria on the orders of Julius Caesar), did not continue through the High Middle Ages.

▶ Vintage sundial clock (©Shutterstock.com)

▼ **1:** One of the earliest and most refined designs of the Egyptian water-clock, dating from the 14th century B.C.

2: The solarium was the perfected Roman version of the sundial.

3: The klepsydra, ancient water-clock, was also a very sought-after decorative item.

4: A drawing of the spectacular Ktesibios water-clock dating from the 2nd century B.C. A little statue was shifted by a float mechanism where a water-wheel turned the column with the time scale. A special device bounced a stone into a little dish and thus sounded the hours.

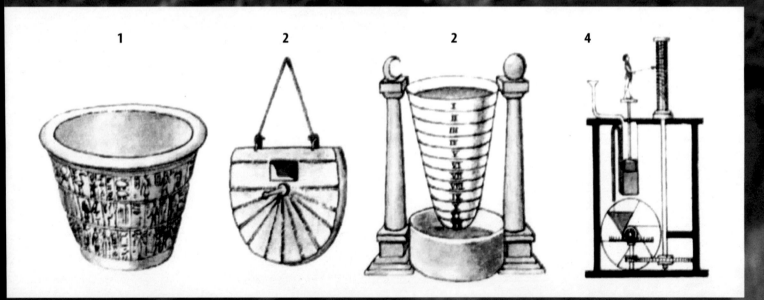

PRECISELY SPEAKING

"THE QUEST FOR HARMONY IS NEVER-ENDING." | BENJAMIN CLEMENTINE, MUSICAL ARTIST, WEARS THE VACHERON CONSTANTIN FIFTYSIX.

 | ONE OF NOT MANY.

CONTACT US: +33 1 70 70 20 27

PRECISELY SPEAKING

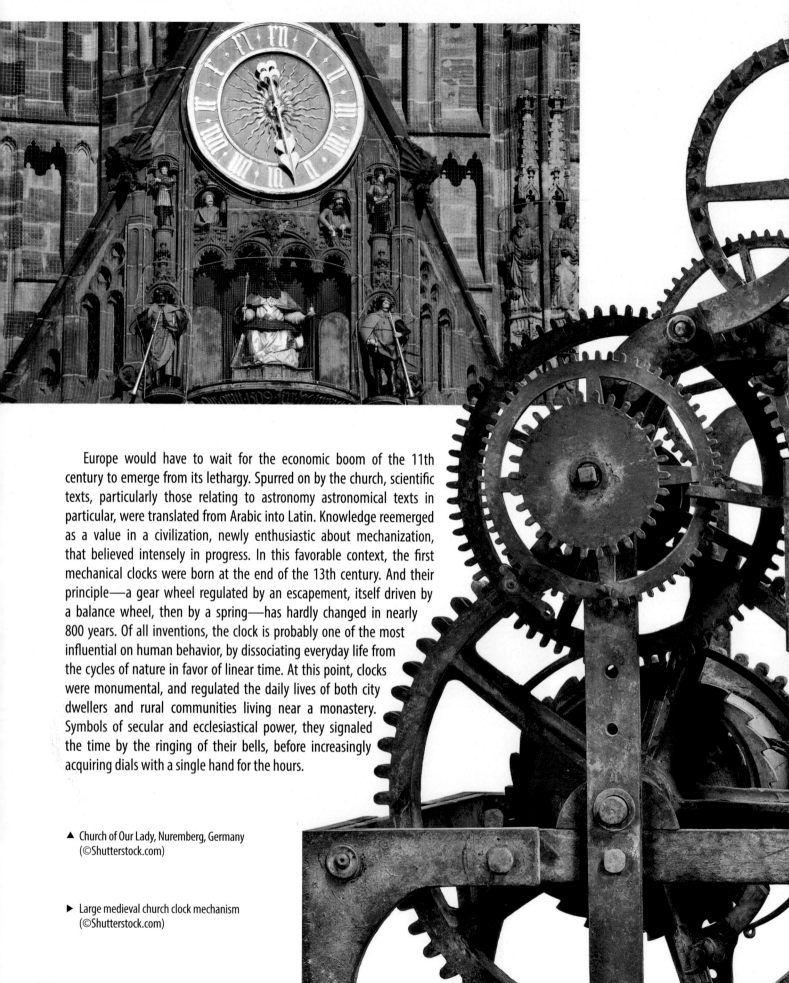

Europe would have to wait for the economic boom of the 11th century to emerge from its lethargy. Spurred on by the church, scientific texts, particularly those relating to astronomy astronomical texts in particular, were translated from Arabic into Latin. Knowledge reemerged as a value in a civilization, newly enthusiastic about mechanization, that believed intensely in progress. In this favorable context, the first mechanical clocks were born at the end of the 13th century. And their principle—a gear wheel regulated by an escapement, itself driven by a balance wheel, then by a spring—has hardly changed in nearly 800 years. Of all inventions, the clock is probably one of the most influential on human behavior, by dissociating everyday life from the cycles of nature in favor of linear time. At this point, clocks were monumental, and regulated the daily lives of both city dwellers and rural communities living near a monastery. Symbols of secular and ecclesiastical power, they signaled the time by the ringing of their bells, before increasingly acquiring dials with a single hand for the hours.

▲ Church of Our Lady, Nuremberg, Germany
(©Shutterstock.com)

▶ Large medieval church clock mechanism
(©Shutterstock.com)

PRECISELY SPEAKING

▶ 18th-century table clock
(©Shutterstock.com)

THE INVENTION OF PRECISION

Needless to say, precision was not the primary quality of these monumental clocks. Equipped with a verge escapement (a horizontal balance coupled with a crown wheel), they could run slow or fast by up to two hours per day, making the presence of a minutes hand completely incongruous. To set the time, it was customary to refer to the good old sundial. Faced with the difficulties of creating more reliable mechanisms, watchmakers also set to work on an equally crucial task: the miniaturization of mechanisms, a guarantee of independence from "public" time. The progress made in metallurgy from the 15th century onwards and the evolution of horological mastery would thus lead to the creation of the first balance springs, brilliant substitutes for motor weights. Now portable—and decorative—the indoor clock made its appearance, before the so-called "haute époque" watch would in turn become the symbol of a new way of life. To compensate for their notorious imprecision, these watches took on diverse and varied forms, or bore rich adornments as decoration. More important than the time-telling function was the ornamental one; they acted as mechanical jewels during a period when the status outweighed the usefulness of the instrument, especially worn (as the nobility did) at the waist, hung on a necklace, or even embedded in a ring or a dagger handle.

▼ Modern reconstruction of de Dondi's astronomical clock. It took de Dondi from 1348 to 1364 to design and construct the clock.

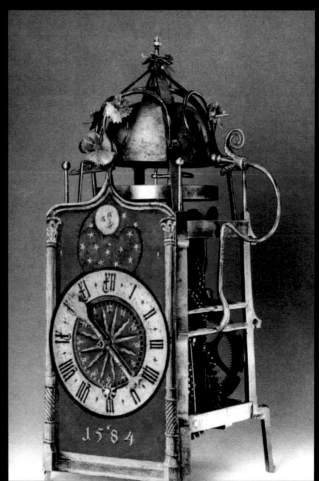

▼ Early clocks were forged by blacksmiths, and typically made of iron. This iron wall clock, circa 1584, features a gothic dial and movable moon plate to show phases of the moon.

PRECISELY SPEAKING

▲ Galileo Galilee

These frivolous concerns did not, however, seal the fate of watchmaking. Particularly during the 16th century, great minds were at work. From the 16th century onwards, watchmaking would benefit from foundational inventions by scientists with wide-ranging knowledge such as Galileo or Huygens, to mention just two. Galileo (1564-1642) was the first to draw practical lessons from the work of Nicolas Copernicus on heliocentrism. At the age of nineteen, around 1583, he discovered the isochronism of the oscillating pendulum: the concept that the duration of oscillations of a pendulum is independent of its amplitude. Thanks to this discovery of constant and regular pendulum oscillations, Galileo provided watchmakers with a reliable and easily reproducible standard, the basis of work that would enable watchmaking precision to take a giant leap.

▶ Created around the year 1600, this ivory portable sundial is decorated with garnets and emeralds.

▲ Replica of a medieval astrolabe (©Shutterstock.com)

1888

MADE OF LUCERNE

Carl F. Bucherer

LUCERNE 1888

MANERO TOURBILLON DOUBLE PERIPHERAL
FLOATING TOURBILLON | 18 K ROSE GOLD

carl-f-bucherer.com

PRECISELY SPEAKING

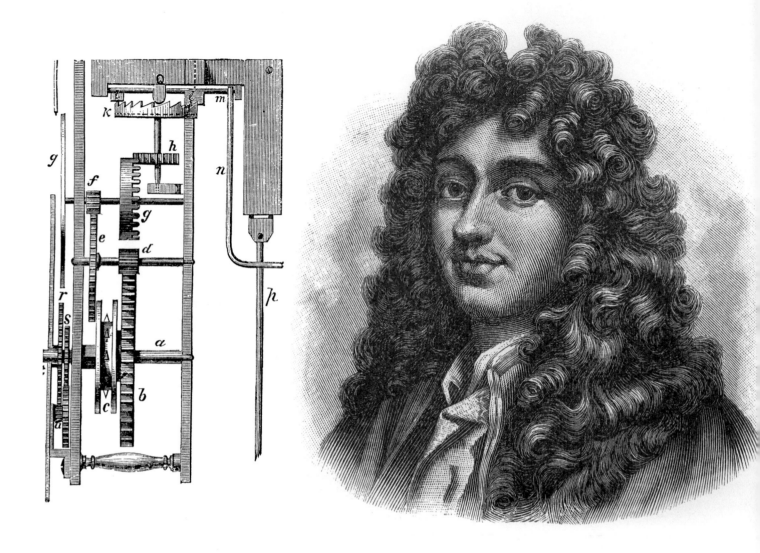

▲ *above left*:
Huygens Clock or Pendulum Clock, vintage engraving.
(©Shutterstock.com)

▲ *above right*:
Christiaan Huygens
(©Shutterstock.com)

Dutch physicist and astronomer Christiaan Huygens (1629-1695) would take up the baton. Following the work of Galileo, he created a mechanical clock still driven by the action of weight and spring, but whose escapement was now regulated by a pendulum. In his book The Mastery of Time, Dominique Fléchon explained how a pendulum, hanging freely from a wire and thus isolated from the gearwheel, communicates an alternative movement to the escapement wheel, whose gears are now subject to shocks of constant intensity. It is also to Christiaan Huygens that we owe the first sprung-balance watch, a truly revolutionary concept for the time. The regulating organ was at that point formed of a balance, a small mechanical flywheel, coupled to a fine spiral-wound steel spring which imparts a back-and-forth movement under the impulse of the escapement. The balance-spring would fulfill for the watch the same regulating function as the pendulum for the clock. In this sense, the date of January 30, 1675 would mark a milestone in watchmaking history. It was on this date that Christiaan Huygens presented his sprung balance to the Royal Society of France. This invention reduced the time lost each day by the best models on the market from a good half hour to four or five minutes. With this level of precision, watches became sufficiently reliable instruments to be of interest to physicists and astronomers in their research. This progress would justify the presence of the minute hand, which became widespread after 1700.

PRECISELY SPEAKING

◄ This Renaissance table clock, circa 1640, is crafted of a bronze lion and small dog. As the clock works, both the lion and the dog move their eyes in time with the clock. As the hour strikes, the lion opens its mouth and moves its wings.

◄▲ Interior and exterior of enamel timepiece from the 17th century. Enamel work became an important decorative function of watches and clocks.

PRECISELY SPEAKING

THE NAVY STICKS ITS OAR IN

The work of Huygens, the father of scientific watchmaking, and the "precision revolution" on which these scientists embarked paved the way for new investigations that would soon lead to the resolution of the calculation of longitude at sea. Seafarers played a central role in enhancing watchmaking precision—and not just for the hell of it. As is often the case, the economic stakes dictated the progress made in this area. By the 15th century, the conquest of the seas was supposed to open new sea routes to lands of plenty, synonymous with extended powers and new wealth for governments who launched their squadrons into the adventure. Coastal navigation was jettisoned, even with the knowledge that navigation based on observation of the stars could easily end in tragic shipwrecks. At the time, although a ship's latitude had long since been calculable using the position of the sun at noon or the pole star at night, its longitude posed a problem in the absence of reliable chronometers to accurately calculate the time traveled by a ship between two points.

▶ Vintage measuring instruments for navigation
(Alevtina Vyacheslav ©Shutterstock.com)

▼ Vintage sextant with compass
(©Shutterstock.com)

PRECISELY SPEAKING

PRECISELY SPEAKING

To make up for this, sailors used to mount sundials on compasses. But the imprecision of this approach and the associated shipwrecks—including that of four British warships on the coast of Cornwall in 1707 causing the death of 2,000 sailors—would precipitate things. In 1714, the British Parliament enacted the Longitude Act offering a reward of 20,000 pounds, a colossal sum for the time, to whomever could develop a method to determine the longitude to a half degree for a trip from Great Britain to the West Indies, a margin of error of about 30 nautical miles (~ 56 km) after 40 days of navigation. The Académie des Sciences in Paris did the same four years later. With watchmakers as their proxies, England and France then engaged in a merciless struggle, continually outdoing each other in ingenious refinements to their marine chronometers... not without a few episodes that today we would call industrial espionage. Christiaan Huygens himself sought this Holy Grail but without success: his balance-spring system was much too sensitive to temperature differences to ensure a sufficiently reliable chronometry.

▶ Old compass, astrolabe on vintage map
 (©Shutterstock.com)

PRECISELY SPEAKING

Constellation ★ *Manhattan*

18K Sedna™ gold with diamond-paved bezel, mother-of-pearl dial and diamond hour markers.

MASTER CHRONOMETER CERTIFIED.

Ω
OMEGA

▶ UK postage stamp circa 1993 depicts Marine Chronometer by inventor John Harrison. (Chris Dorney ©Shutterstock.com)

▼ John Harrison (1693-1776) (©Shutterstock.com)

The challenge was daunting. The best clocks of the time showed daily variations of several minutes, and the specifications for marine chronometers as stipulated by the Longitude Act pegged the acceptable lag at little more than one second per day, for instruments tossed by the waves in environments that would grow progressively more humid and hot as the journey went on. Many master watchmakers took up the gauntlet, trying to solve the era's equivalent of "squaring the circle": the creation of a high-performance escapement, coupled with a balance unaffected by temperature variations, and with the entire instrument immunized against the ship's constant movements. The solution was to come from an English carpenter-watchmaker, John Harrison, who solved the problem for the first time in 1734 with a huge marine chronometer weighing 32 kg. Difficult to reproduce, it was perfected over the years; John Harrison and his son William finally won the prize in 1772 with the H4, an extraordinary marine chronometer, resembling a large pocket watch with a diameter of 13 cm and weighing 1.5 kg. After crossing the Atlantic, the H4 experienced a lag of only 39.2 seconds in 47 days of navigation, or 0.834 seconds per day, for a total distance difference of 9.8 miles (~ 18 km).

▲ Kirov watch for maritime uses
(Matveychuk Anatoliy ©Shutterstock.com)

PRECISELY SPEAKING

▲ Elgin's Lincoln Plant

PRECISELY SPEAKING

LESSON FROM AMERICA

This progress made in the world of nautical chronometry would certainly not have been possible outside the restricted world of the workshops of master watchmakers, who practically crafted watches one at a time. Admittedly, the models had to be designed with reproducibility in mind, but production was limited to a few dozen timepieces at most, given the difficulties in shaping these high-precision instruments, and their concomitant prohibitive cost. In the case of watches, given the need to meet the growing demand in large urban centers, production had already undergone a technical and social division of labor. Given that the manufacture of the multiple components of a timepiece required more time than its assembly, these two functions had already undergone parallel developments. Home-based workers—manufacturers of components, dials and hands, draftsmen, case builders—soon to be grouped into larger structures, supplied shopkeepers who, in turn, finished and adjusted parts as well as marketing the completed watches. This was also the beginning of mechanization, which enabled watchmakers to increase production from two base movements per worker per month to four or more per worker per week. Unusually for the time, in 1780, Frédérique Japy's workshop in the Swiss Jura managed to deliver more than 20,000 "blancs-roulants" (completed movements without escapements) to the watchmakers of Neuchâtel. In comparison, the factory founded by Voltaire in Ferney had an annual production of some 4,000 units and Geneva as a whole 40,000.

▲ Pilots receiving Hamilton watches

◀ Hamilton Company in Lancaster

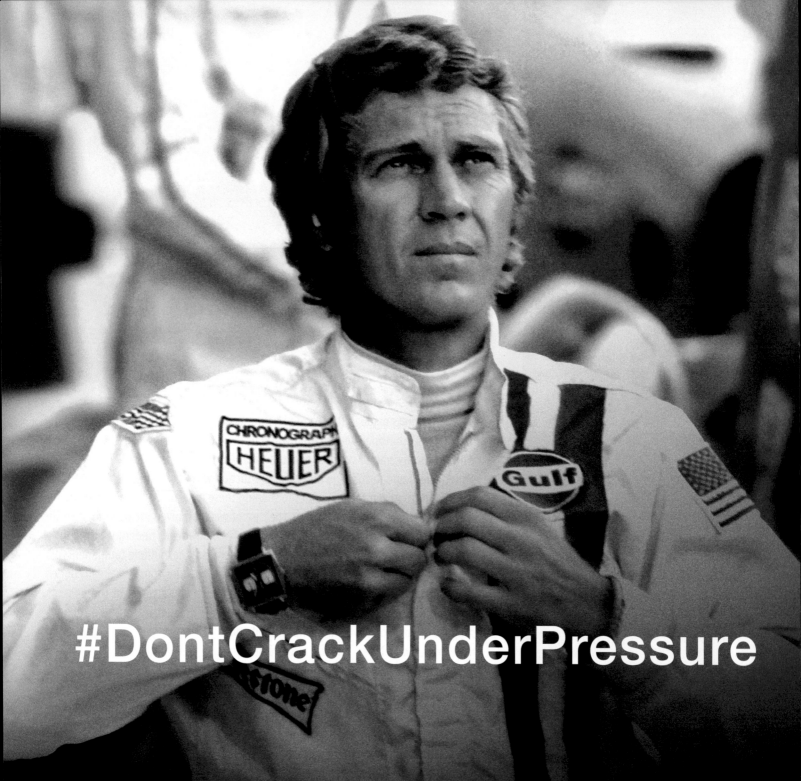

SWISS AVANT-GARDE SINCE 1860

MONACO CALIBRE 11 AUTOMATIC CHRONOGRAPH

Steve McQueen's legacy is timeless. More than an actor, more than a pilot, he became a legend. Like TAG Heuer, he defined himself beyond standards and never cracked under pressure.

PRECISELY SPEAKING

But Switzerland was certainly not at the forefront of this more efficient production system. In the midst of the Industrial Revolution, triggered by the steam engine (whose first functional example dates back to 1712), it was even seriously lagging behind other countries, particularly because of the resistance from the guilds, fiercely attached to their independence and manual work. However, nothing of the kind existed in the United States, which welcomed Pierre-Frédéric Ingold (1787-1878) like a messiah. This Swiss engineer, who had developed machines to cut out components, split and finish wheels, and even turn watch cases, had hit a wall in the Old World. The American manufacturers of Boston, however, rushed to adopt his machines and methods. In just a few decades, companies like Howard, Hamilton, Elgin, Illinois Watch Co., Dueber-Hampden, Burlington, and Waltham had built reputations both nationally and internationally. The example of Waltham is most edifying. Founded in 1850, in the midst of the Industrial Revolution, the Manufacture quickly pioneered production lines based on the principle of division of labor and movement design with completely interchangeable components. In addition, solid industrial logic rejected the idea of working with trades scattered hither and yon—Waltham brought all of its activities under one roof, including the production of its tools and its own alloys.

▲ Admiral E. Byrd pocket watch for Antarctica Expedition (1928-1930)

PRECISELY SPEAKING

"Off duty — a dress watch"

The Elgin
B. W. Raymond — 21 Jewels

There's nothing wrong with a railroad man's eye—when it comes to beauty

If you mention that famous Elgin Railroad Watch—the B. W. Raymond—to a veteran railroad man, you're apt to tap a geyser of enthusiasm about its trustworthiness, its stamina, its unflagging accuracy.

He'll tell you of its remarkable Invar Balance principle — which gives this watch the closest possible rating through all degrees of temperature.

He'll tell you of its eight adjustments, five of them to position, which keep this watch in perfect service under the most drastic ordeals of railroad service.

But he may not take time to tell you that the B. W. Raymond is a winner *in looks* as well as efficiency. He'd rather let you see that for yourself.

For a railroad man isn't given to talking much about the artistic side of things. He lives a life of practicality. But there's nothing the matter with his artistic eye. He takes a quiet joy in the fact that his Elgin is *styled to the minute* as well as timed to the second.

What a peach of a watch! Not clumsy or heavy as you might expect such a stalwart time-keeper must be. But lithe, trim and graceful of line. A watch that combines maximum service efficiency with maximum watch elegance.

ELGIN
THE WATCH WORD FOR ELEGANCE AND EFFICIENCY
ELGIN NATIONAL WATCH COMPANY . . . . . . ELGIN, U. S.

PRECISELY SPEAKING

▲ Waltham manufacturing facilities

Though Waltham stood out for its production process, the design of its calibers was equally noteworthy. The Manufacture was among the first to adorn its barrels, and to study friction reduction methods and even new structures, to optimize the positioning of components and reduce their number. The quest for precision went hand in hand with this watchmaking philosophy, leading Waltham to build its own internal astronomical observatory and in the process create its own chronometric standard, more stringent than those in force at the time. The Philadelphia World's Fair in 1876 was the apotheosis of the American horologer, which installed a complete production line on site. In six days of work, at the rate of 10 hours a day, to visitors' amazement, Waltham would produce more than 2,200 gold or silver watches, offered at half the price of equivalent Swiss models. Swiss watchmakers had more nasty surprises in store that would leave them shocked and stunned. It was a Waltham watch, chosen at random on the Exposition's production line, which earned the gold medal at the first international timekeeping precision competition, organized for the occasion. Switzerland was dismayed at the revelation of its humiliating relative technological backwardness.

©Shutterstock.com

▶ Waltham machines

PRECISELY SPEAKING

COMPETITION

Switzerland was not long in responding. Towards the end of the 19th century, exports of Swiss watches to the United States were dismally low, and with good reason. In addition, problems with supply lines and organization became glaring enough for some Swiss manufacturers returning from the United States to actively campaign for an upgrade of watch production facilities and capabilities. Some Manufactures, such as Longines or Vacheron Constantin, had already established industrial processes that resembled American "factories," but the profession as a whole remained largely dependent on tradition. The watchmakers of Switzerland turned to defending their interests, finding a sympathetic ear with the Swiss government. In November 1880, the first federal law concerning the protection of trademarks came into effect. In its wake, the authorities in Geneva—a center of Swiss watchmaking that was then in the process of building a reputation for technical parts—enacted a law in 1886 establishing the Geneva Seal (Poinçon de Genève) as a label of origin and quality reserved for watches produced in the canton that demonstrated "all the qualities of production likely to ensure regular and durable functioning." At the same time, the observatories of Neuchâtel and Geneva took on the mission of stimulating the watch industry through chronometry competitions, while providing manufacturers with individually issued certificates of precision for their products.

Big Bang Unico Blue Magic.
Case in vibrantly-coloured blue ceramic. In-house UNICO
chronograph movement. Interchangeable strap using
patented One-Click system. Limited edition of 500 pieces.

PRECISELY SPEAKING

The Geneva Observatory, built in 1772 and equipped with a chronometry service at the end of that century, was thus pressed into service to support the development of Swiss watch companies. Its first chronometry contests, organized without precise rules and regulations, evolved into more formal competitions, inaugurated in 1872. The Neuchâtel Observatory did the same. As explained in the 2014 traveling exhibition "Omega Chronometer Watches,""the first chronometry competitions were launched at the end of the 19th century. They were mainly organized by the most famous observatories of the time, located in Geneva, Neuchâtel, Besançon or London (Kew). The purpose of these competitions was to designate the best manufactory or master watchmaker among all the participants. All were separated according to a single criterion: precision. The watch companies committed a lot of resources and time to these competitions. This quest for perfection had a scope that went far beyond its scientific aspect. A victory could have invaluable commercial effects because it allowed one watchmaker to definitively establish its superiority with respect to all its competitors."

▼ L.U.C movement
(Chopard)

PRECISELY SPEAKING

▲ Neuchâtel Observatory

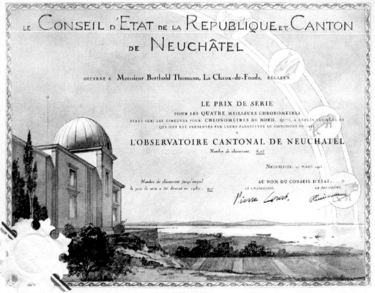

ROGER DUBUIS
DARE TO BE RARE

The National Physical Laboratory,

RATING DEPARTMENT.

I hereby Certify

That a CLASS A KEW CERTIFICATE

has been issued to *The Rolex Watch Co.*
London & Bienne:-

for the *Keyless crystal 11 line bracelet* Watch

No. *492282*, which was submitted to a Trial at this

Institution extending over 45 days from *June 1* to

July 15 1914, and the results of its performance were

such as to entitle it to this Certificate, in accordance with the

Regulations for the issue of Watch-rate Certificates approved

by The National Physical Laboratory Committee of the

R. T. Glazebrook

DIRECTOR.

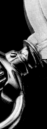

▶ **Pocket watch by A. Lange & Söhne with center second from 1896.**
(A. Lange & Söhne)

All the effort expended in producing and fine-tuning these timekeeping instruments would, however, be thrown into question by the arrival of the wristwatch. Just about every aspect had to be completely rethought, taking into account the new miniature size as well as functionality. And since these objects were now also exposed to the vagaries of everyday life, the level of difficulty was compounded by concerns about water resistance, shock protection, and the fragility of the protective glass. The watchmakers had little choice but to adapt. First seen as an eccentric fad, the wristwatch would become indispensable over the early decades of the 20th century, concurrently with a wave of innovations to make these pieces affordable, useful, and reliable. As recalled by The Mastery of Time, Rolex received the first official rating certificate for a wristwatch in 1910, and in 1914, the English Kew Observatory granted Rolex the first 'Class A' certificate for a wristwatch. The brand had finally proven that a wristwatch could reach the precision of a pocket watch.

▼ **Oyster Perpetual Datejust**
(Rolex)

PRECISELY SPEAKING

▶ **Santos**
(Cartier)

THE SUPREME CHRONOGRAPH

From that point on, with precision rediscovered in this new format—and thus within the grasp of horologers converted to the gospel of the wristwatch—many models would fulfill the specific needs of functionality. Consider the first pilot's watches, including the now famous Santos de Cartier, the IWC Pilot's Watches or the Pilot models from Zenith, Breitling's Navitimer, and Type 20 from Breguet.

▶ **Aviator Double Chronograph**
(IWC)

PRECISELY SPEAKING

◄ **Pilot Type 20 Extra Special Chronograph** (Zenith)

▲ **Navitimer** (Breitling)

▲ **Type XX Chrono** (Breguet)

PARIS

CIRCLE « LA PETITE »

Dotée d'un boîtier ultra-plat de 45 ou 52 millimètres de diamètre pour 6 millimètres d'épaisseur pour la première et 7 millimètres pour l'autre, cette montre d'une extrême convexité épouse parfaitement le poignet de toutes les dames. Son boîtier se décline en or blanc, rose ou jaune avec lunette sertie ou entièrement serti. Son cadran au galbe parfait et à la finition poli-miroir ou satinée se définit au choix avec index en appliques, index sertis ou encore entièrement serti neige.

With an ultra-flat case of 45 or 52 millimetres in diameter and 6 millimetres thickness for the first and 7 millimeters for the other, this watch literally envelops women's wrist thanks to its impressive convexity. This case is available in white gold, rose gold and yellow gold, with a bezel set or full set. Its dial with a perfect curve and with a gold mirror or a gold matte aspect is available with simple markers, jewel-set markers or completely set.

PRECISELY SPEAKING

▲ **Fifty Fathoms** (Blancpain)

Scuba diving was also becoming a major challenge for watchmakers competing for the wrists of divers, like Blancpain with its Fifty Fathoms, Omega with its Seamaster, Rolex with its Submariner, or Jaeger-LeCoultre with its Memovox Deep Sea and Girard-Perregaux with its Sea Hawk.

▶ **Seamaster SE Planet Ocean Deep Black** (Omega)

▶ **Seamaster** (Omega)

PRECISELY SPEAKING

▼ **Submariner** (Rolex)

▶ **Deep Sea Alarm Automatic** (Jaeger-Le Coultre)

▶ **Sea Hawk 3000** (Girard-Perregaux)

▲ **Oyster Perpetual Explorer II** (Rolex)

PRECISELY SPEAKING

▶ **Oyster Perpetual Explorer II**
(Rolex)

PRECISELY SPEAKING

For explorers, timekeeping instruments were just as indispensable. These adventurers might choose to wear a Rolex Explorer, Tudor Oyster Prince, or maybe Doxa Sub. In this context, the watch was an instrument, and an essential professional accessory, even a life-saving one.

◀ **SUB 300 Professional 2017**
(Doxa)

▶ **Oyster Prince**
(Tudor)

PRECISELY SPEAKING

▶ **Speedmaster**
First Omega in space
(Omega)

PRECISELY SPEAKING

Everyone still remembers the story of the Apollo 13 mission in 1970—an aborted mission whose crew owes its life to Commander Lovell's Omega Speedmaster, which allowed him to time the trajectory adjustments of the module to ensure a safe re-entry into the atmosphere. Plainly, the chronograph had fully earned its stripes by then. In addition, throughout the first half of the 20th century, competitive sports became more popular and widespread, requiring professional instruments of indisputable reliability. Precision was best expressed by the ability of watches to measure short times.

▶ **Speedmaster**
Gemini 4 mission, June 1965
(Omega)

GUY ELLIA

PARIS

« JUMBO CHRONO»

Mises en valeur par un imposant boîtier de 50 x 11,5 mm , les finitions et les caractéristiques de la Jumbo Chrono font de cette pièce unique un modèle d'élégance et de subtilité. Agrémentés d'un fond glace Saphir, les quatre cadrans indicateurs sont animés par un mouvement automatique chronographe avec roue à colonne, conçu par "Manufacture Blancpain", et offrant une réserve de marche de 45 heures. Ses ponts, son cadran et sa lunette en or gris pvd noir microbillé sertie de 323 brillants pour 7.93 carats et sa boucle déployante sont assortis d'un bracelet Alligator.

Highlighted by an impressive case 50 x 11.5 mm, the finishes and the features of the Jumbo Chrono make of this unique piece a model of elegance and subtlety. Orned with a sapphire crystal back case, the four dials are driven by an automatic movement, chronograph with column wheel, designed by "Manufacture Blancpain", and offering a power reserve of 45 hours. Its bridges, its dial and its bezel in black matte pvd treatment set with 323 brilliants for 7.93 carats and its deployant clasp embellished by an alligator strap

PRECISELY SPEAKING

Short-term time measurement is a problem that was an early concern for horologers, including Louis Moinet, to whom we owe the "compteur de tierces" (thirds counter) in 1816, a mechanism precise to the 60th of a second thanks to a caliber beating at 216,000 vph (30 Hz). This feat was enough for Louis Moinet to be recognized today as the father of the chronograph and a high-frequency pioneer—especially when we consider that today, the most-used frequency in watchmaking is 4 Hz, or 28,000 vph. Initially used in astronomical observations, chronographs have clearly become essential instruments in the world of sports. And since athletics have occupied a special place in all civilizations since time immemorial, the progress of watchmaking has been partly dictated by the quest for precision dictated by the demands of sporting competition.

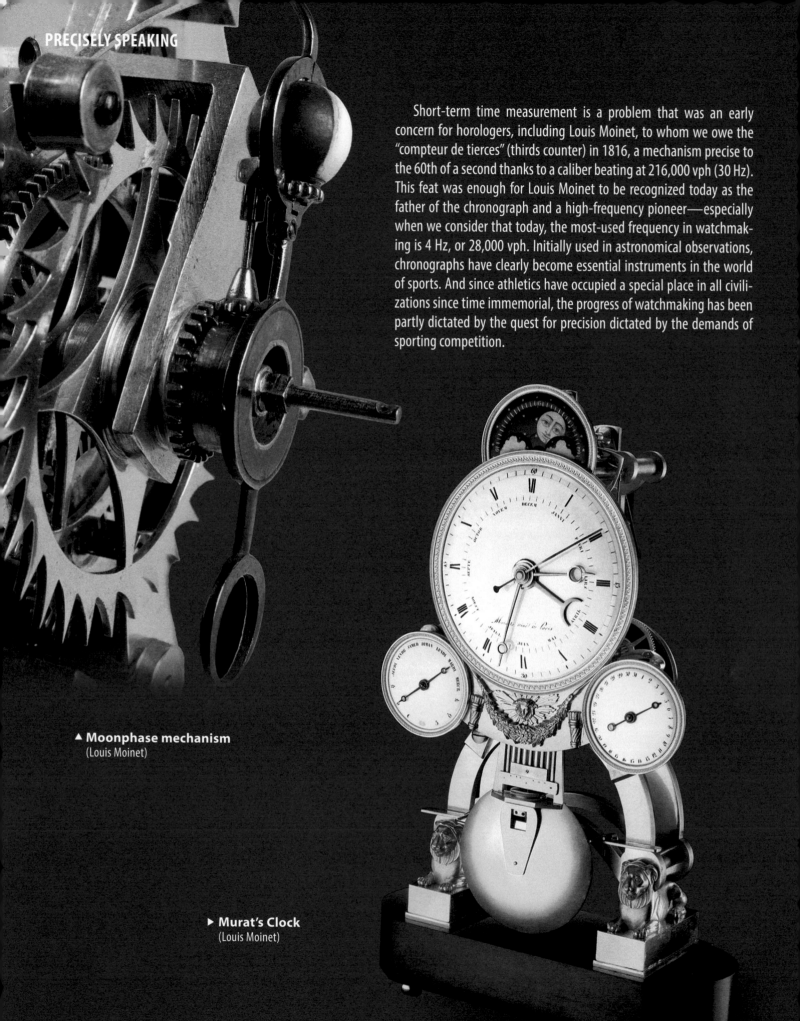

▲ **Moonphase mechanism**
(Louis Moinet)

▶ **Murat's Clock**
(Louis Moinet)

PRECISELY SPEAKING

▲ **Chronograph Pocketwatch**
(Louis Moinet)

PRECISELY SPEAKING

▲ 1500m Men's semifinals run in the Rio 2016 Olympics Games (CP DC Press: ©Shutterstock.com)

PRECISELY SPEAKING

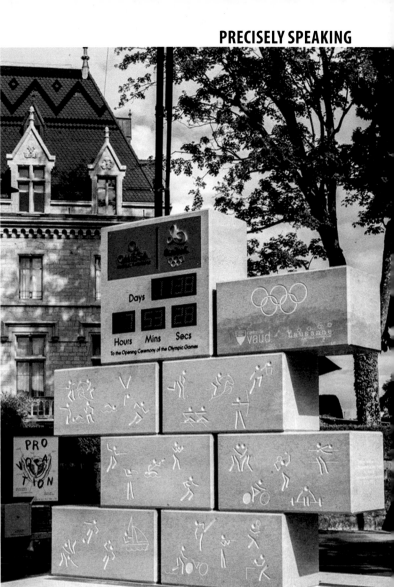

▲ Omega has served as the official timekeeper of the Olympics many times.

Some houses would stand out by developing the features of what would become the modern chronograph wristwatch, equipped with a split-seconds feature thanks to Patek Philippe in 1927, with two pushers following the invention patented by Breitling in 1933, with a flyback function thanks to Longines in 1936, and finally a regatta countdown, released by JeanRichard in 1961. Constant improvement of the calibers was also on the agenda, especially at Omega, the serial Olympic timekeeper, with its legendary Speedmaster first equipped with the manual-winding 321 movement, then replaced by the 861 and its improved versions.

◀ **Omega trilogy** (*from left to right*)
1957 Seamaster, Railmaster, Speedmaster
(Omega)

PRECISELY SPEAKING

◄ Carrera
(TAG Heuer)

At Zenith, the legendary El Primero would go down in history as the first mass-produced chronograph accurate to a tenth of a second. As for TAG Heuer (or just "Heuer" at the time), to whom we owe the first automatic chronograph in 1969, its Carrera collection earned the brand a place in the chronograph world. In all, no less than 11 generations of Carrera have emerged since 1963, each iteration bringing evolutions—and sometimes revolutions.

◄ El Primero Original
(Zenith)

IWC PORTUGIESER.
THE LEGEND AMONG ICONS.

Portugieser Automatic. Ref. 5007: Where you go is entirely up to you. After all, it's your boat. So if you're in the mood for sailing round the world, away you go. The pocket watch movement of your Portugieser Automatic with its seven-day power reserve and Pellaton winding will be only too happy to oblige. Because its precision will always navigate you reliably to your destination. Assuming you always take it with you. **IWC. ENGINEERED FOR MEN.**

Mechanical IWC-manufactured movement 52010 calibre · Pellaton automatic winding · 7-day power reserve · Power reserve display · Date display · Small hacking seconds · Breguet spring · Rotor with 18-carat gold medallion · Sapphire glass, convex, antireflective coating on both sides · See-through sapphire-glass back · Water-resistant 3 bar · Diameter 42.3 mm · Stainless steel

IWC Schaffhausen, Switzerland · www.iwc.com

FOLLOW US ON:

IWC
SCHAFFHAUSEN

PRECISELY SPEAKING

◄ ▲ **Carrera Mikrogirder** (TAG Heuer)

The recent Carrera Mikrogirder's seconds hand performs 20 turns around the dial per second, making it the fastest mechanical watch ever made, able to time increments of 1/2000th of a second (1,000 Hz). In the category of extraordinary contemporary chronographs, we must mention A. Lange & Söhne's 2018 Triple Split, which can store an intermediate or comparative time in its memory for 12 hours.

▶ **Triple Split**
(A. Lange & Söhne)

PRECISELY SPEAKING

THE QUARTZ WAVE

The advent of quartz, however, would set the record straight on accuracy. The first harbinger came, again, from the United States in 1960 with the Accutron (accuracy through electronics), a wristwatch produced by the American Bulova Manufacture and based on an invention by Swiss engineer Max Hetzel. His idea was to replace the oscillations of the sprung balance or the pendulum by the vibrations of a tuning fork to regulate the rhythm of the movement. Alarm bells rang throughout the industry. The Bulova Accutron offered six times the accuracy of mechanical watches with a one-year autonomy. Swiss watchmakers rushed to prepare a counter-attack. Concerned by the danger, the Federation of the Swiss Watch Industry decided to set up a new research institute to organize a counter-offensive, well aware that the tuning fork system was protected by patent. The solution had to involve quartz and the miniaturization of existing electronic clocks, while large enough to make it clear that actually incorporating these mechanisms within wristwatches was a utopian pipedream.

It took five years for the Centre Electronique Horloger to produce two working prototypes, Beta 1 and Beta 2. The same year, in 1967, a series of 11 movements of the two versions was presented at the Chronometry Competition at Neuchâtel Observatory. Jackpot! They were twelve times more accurate than a mechanical chronometer and six times more accurate than tuning-fork watches. But the triumph would soon curdle into bitter disappointment. Seiko researchers, who were working on a similar project, developed the Astron 35SQ Quartz, with an accuracy of one minute per year, released on Christmas Day 1969. As Japanese watchmakers had been the first to invest in this promising new technology, much more precise and ridiculously cheap, the quartz tidal wave flooded the world markets. Swiss horologers, clinging to the idea of the electronic watch as an exceptional piece reserved for the elite—an extension of the mechanical tradition—totally missed the boat. As a result, the quartz crisis lasted 15 years. Between 1970 and 1985, the Swiss watch industry shrank by two thirds—both in terms of jobs (from 90,000 to 30,000) and companies (from 1,800 to 600).

PRECISELY SPEAKING

◂ **Seiko 1969 Quartz Astron**
(Seiko)

▾ A Seiko Astron, the world's first quartz watch, on displayed at the Seiko Watch Museum inside Seiko Dream Square. (Ned Snowman ©Shutterstock.com)

PRECISELY SPEAKING

▶ **Manero Flyback**
(Carl F. Bucherer)

◀ **Grande Lange 1 Moon Phase**
(A. Lange & Söhne)

As we know today, the era of mechanical watch was far from over. In the 1980s, the first wristwatch auctions helped revive watchmaking culture. The release of new technical publications, as well as those intended for a broader audience, and the proliferation of watchmaking magazines, gave new life to a watchmaking industry that boasted more enthusiasts than ever. In addition, the Internet was rapidly becoming a powerful method of sharing knowledge. Traditionally styled timepieces, which had seemed incredibly old fashioned and obsolete, once again became the symbols of entrancing, age-old knowledge. As opposed to the electronic watch—a perishable gadget sensitive to fashion trends—its mechanical counterpart presents itself as a lifetime "companion", immutable and lasting, which must be cared for as a family heirloom. As digital technology has entered the arena, watchmakers can still compete comfortably with electronics, using traditional techniques—or at least continue to stoke the passion of their fans with models of irreproachable reliability. Any new triumph of the mechanical watch is inevitably bound up with its precision.

IWC PORTUGIESER.
THE LEGEND AMONG ICONS.

Portugieser Perpetual Calendar. Ref. 5033: The daring expeditions of the Portuguese seafarers held out the promise of everlasting glory. A worthy legacy of this heroic epoch is the Portugieser Perpetual Calendar. Timelessly elegant, it features trailblazing technology that includes a 7-day automatic movement with Pellaton winding and a power reserve display showing the date until 2499. So converted into human lifetimes, this model could be working on its legendary status for eternity. **IWC. ENGINEERED FOR MEN.**

Mechanical IWC-manufactured movement 52610 calibre · Pellaton automatic winding · 7-day power reserve · Power reserve display · Perpetual calendar with displays for the date, day, month, year in four digits and perpetual moon phase · Small hacking seconds · Breguet spring · Rotor in 18-carat red gold · Water-resistant 3 bar · Diameter 44.2 mm · 18-carat red gold · Alligator leather strap by Santoni

Discover the Portugieser Collection

IWC Schaffhausen, Switzerland · www.iwc.com

FOLLOW US ON:

IWC
SCHAFFHAUSEN

▲ Abraham-Louis Breguet

THE ART OF THE TOURBILLON

This quest for "the correct time" enjoyed the significant advance of being able to build on previous research, at least in one particular field: that of the tourbillon escapement, intimately associated with functioning regularity since it is supposed to counter the effects of earth's gravity on watch mechanisms. For a long time, this invention of Abraham-Louis Breguet, patented in 1801, was considered the pinnacle of the watchmaking art, notably because of its complexity.

▲ Abraham-Louis Breguet's patent for the tourbillon

▲ Abraham-Louis Breguet's tourbillon design

▲ Breguet Manufacture   ▲ Creating a tourbillon

PRECISELY SPEAKING

▸ **Octo Finissimo Tourbillon** (Bulgari)

As a result, there were few houses with the necessary technical expertise to produce it on an industrial scale. This scarcity is a thing of the past. To demonstrate their technical mastery, many watchmakers have stepped up to the plate, and not necessarily the usual names in the world of complications, see for example Panerai with its ingenious tourbillon whose carriage rotates perpendicular to the balance spring, or Baume & Mercier with its Clifton 1892 Flying Tourbillon.

▶ **Tourbillon Axe**
(Panerai)

PRECISELY SPEAKING

PRECISELY SPEAKING

Others have approached it as a stylistic exercise, like Franck Muller's Giga Tourbillon Round Skeleton with a tourbillon diameter of 20mm, or Roger Dubuis's Excalibur and Ulysse Nardin's Executive Skeleton Tourbillon, which both play on the concept of extreme transparency.

◄ **Excalibur Squelette**
(Roger Dubuis)

▲ **Excalibur Shootingstar**
(Roger Dubuis)

▼ **Giga-Tourbillon**
(Franck Muller)

◄ **Tourbillon Transparence**
(Ulysse Nardin)

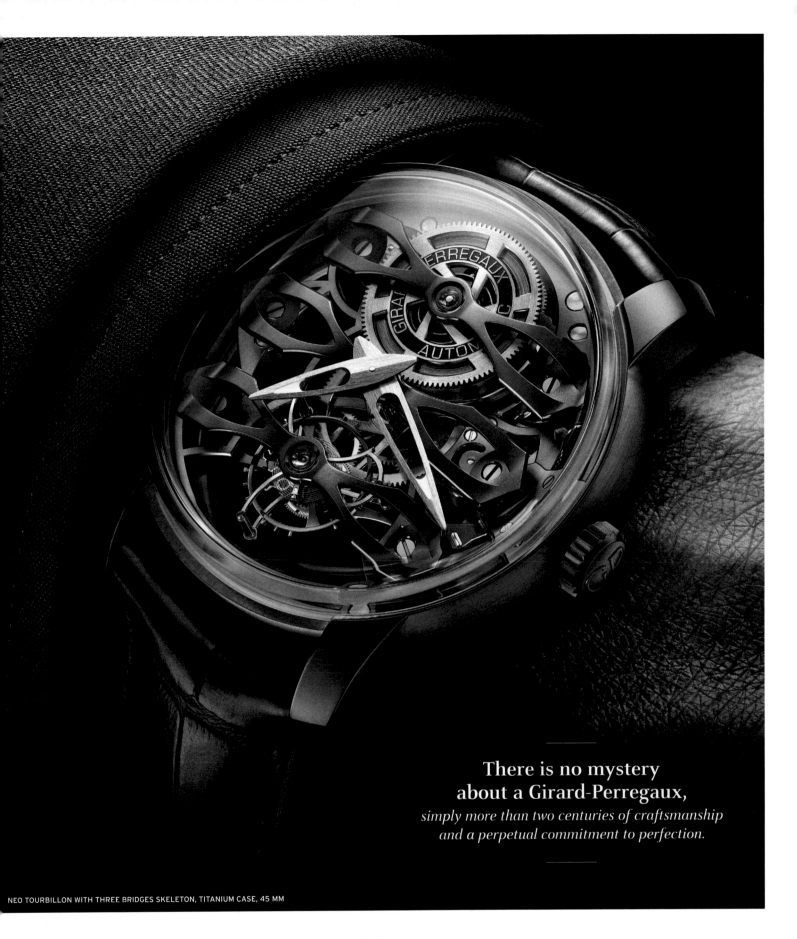

PRECISELY SPEAKING

The truly astounding Octo Finissimo from Bulgari integrates a tourbillon into a movement just 1.95mm thick! Others have ramped up the technical side of this device, such as Harry Winston with the Histoire de Tourbillon collection, Christophe Claret with the X-Trem-1, Jacob & Co. and the Twin Turbo, and Carl F. Bucherer with the Double Peripheral Tourbillon. That's not even counting all the models in which the tourbillon is associated with every imaginable type of horological complication.

▲ **Tourbillon Octo Finissmo**
(Bulgari))

▲ **Histoire de Tourbillon**
(Harry Winston)

◀ **X-TREM**
(Christophe Claret)

PRECISELY SPEAKING

◀ **T3000 in-house caliber**
(Carl F. Bucherer)

▼ **Manero Tourbillon**
(Carl F. Bucherer)

PRECISELY SPEAKING

▲ Le Locle Watchmaking Museunm

However, in terms of accuracy and despite the controversy that still rages about the utility of a tourbillon in a wristwatch (which by definition moves enough on its own to dispense with a mobile escapement), models equipped with such a mechanism are distinguished at the International Chronometry Competition, revived and updated since 2009 by the Locle Watchmaking Museum. Since then, research on the escapement has clearly demonstrated the advantage that could be obtained in terms of reliability, without the tourbillon being the only practical way.

Audemars Piguet has reduced the energy consumption of its Swiss lever escapement via an isochronous system with direct impulse on the balance. Girard-Perregaux has done it with its Constant Escapement system, based on the physical principle of buckling. Not to mention De Bethune and its Isochronic Oscillation System, or the research conducted by Parmigiani on a new flexible joint regulator called Senfine, which promises a months-long power reserve.

▼ **Constant Escapement**
(Girard-Perregaux)

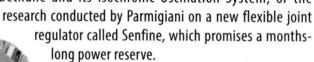

PRECISELY SPEAKING

▶ **Defy**
(Zenith)

The latest example is Zenith's Defy Lab, which is still in extremely limited production, but presented as a completed break with everything that has happened in the field since Christiaan Huygens. The traditional balance-and-spring assembly is replaced by a monolithic organ made of 0.5mm-thick monocrystalline silicon operating without mechanical connections and beating at a frequency of 108,000 vph (15 Hz), three times faster as the legendary El Primero movement. With such a high frequency, the caliber boasts a quantum leap in precision, measured with a daily deviation in rate of just 0.3 seconds!

L.U.C FULL STRIKE

Winner of the Aiguille d'Or at the 2017 Grand Prix d'Horlogerie de Genève, the 42.5 mm-diameter L.U.C Full Strike is Chopard's most sophisticated chiming watch. A unique minute repeater model proudly developed, produced and assembled in our Manufacture. Three patents have been filed for the innovative 533-part L.U.C 08.01-L movement sounding the hours, quarters and minutes on world premiere single-piece transparent crystal gongs.

Chopard

THE ARTISAN OF EMOTIONS – SINCE 1860

New York: Madison Avenue | **Miami:** Bal Harbour Shops, Brickell City Centre | **Costa Mesa:** South Coast Plaza
Las Vegas: Wynn Hotel & Resort | **Houston:** River Oaks District
1-800-CHOPARD   www.chopard.com/us

PRECISELY SPEAKING

Clearly, such a feat achieved in the world of mechanical watches would have been totally impossible without the discoveries made in new materials. This offers a brand-new field of exploration for watchmaking companies inspired by breakthroughs in aeronautics, the space industry and the world of Formula 1. In this respect, Richard Mille is an essential watchmaker, having produced many components of the movement and case in materials formerly unheard of in watchmaking. One example is its RM 50-03 McLaren-F1 tourbillon and split-seconds chronograph that weighs only 40g including the strap, a world record achieved thanks to the use of graphene, a revolutionary nanomaterial 6x lighter than steel but 200x more resistant.

▶ RM 50-03 McLaren-F1 Tourbillon

PRECISELY SPEAKING

Hublot, for whom "The Art of Fusion" is its mission statement, has also distinguished itself in this field, especially with new alloys such as Magic Gold or through the use of carbon fiber, ceramic and tungsten. But as far as precision is concerned, watch manufacturers have been able to claim new performance records because of the emergence of non-ferrous (therefore antimagnetic and self-lubricating) materials, which are produced by shaping, rather than by machining. Among them, silicon has become a must, particularly suitable for escapement systems (levers and lever wheels), and later, the hairspring, in which Ulysse Nardin has been a pioneer since the early 2000s.

◀ **Big-Bang Ferrari Magic Gold**
(Hublot)

THE WAR OF LABELS

With such "arguments" to put forward, watch companies have fully understood the advantage of attacking the standard in force for accuracy, that established by the Official Swiss Chronometer Testing Institute (COSC), which stipulates that to be certified by a battery of tests, the standard watch caliber must not achieve a daily variation in rate within the range of -4 and +6 seconds per day. Year in, year out, the COSC tests nearly 2 million watch movements, mostly mechanical (about a quarter of the annual Swiss production), which are consequently entitled to wear the label on their dial. This is undeniably an added value to the product, given that "Swiss Made" is nothing more than a certificate of origin. As the COSC also explains, "The chronometer certificate is not obtained by chance: it is the symbol of a culture of high quality manufacturing, precision, and exacting standards... The technical value of a certified chronometer compared to a model that does not carry the certification is incalculable." While it is difficult to quantify what the COSC certification means for prices, it is indisputable that it represents a comparative advantage among the most discerning clientele. Among the top-tier brands, out-achieving the standards of the COSC has clearly become a goal, one that is not just attainable, if one considers that the COSC standard dates to 1973, but above all highly beneficial to the desirability of a given watch.

◀ Watches being tested according to the standards of METAS (Federal Office "of Metrology).

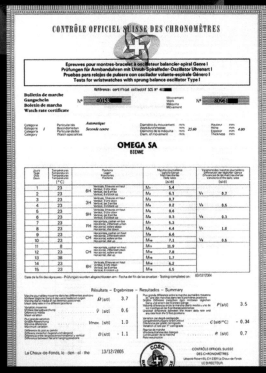

◀ Certificates from the COSC

PRECISELY SPEAKING

Two brands have most prominently taken up the torch, Omega being the first to enter the dance with its Master Chronometer certification. A stroke of genius from the brand, this new label was developed in collaboration with the Institut Fédéral de Métrologie (Federal Institute of Metrology). Any Manufacture is free to submit its watches to the Institute, as long as they are likely to meet its criteria, which are among the most demanding in the industry. Omega has set the bar very high: to obtain the certification, the watches must withstand magnetic fields of 15,000 gauss (the equivalent of the radiation of an MRI device) and demonstrate a functional regularity that does not exceed 0 to +5 seconds per day throughout a series of eight demanding tests.,

PRECISELY SPEAKING

Rolex was bound to respond to such a challenge, duly unveiling its own certification under the name Superlative Chronometer. "The certification applies to the fully assembled watch, after casing the movement, guaranteeing superlative performance on the wrist in terms of precision, power reserve, waterproofness and self-winding," explains Rolex. "The precision of a Rolex Superlative Chronometer after casing is of the order of −2/+2 seconds per day, or more than twice that required of an official chronometer."

▲ **Oyster Perpetual Superlative Chronometer** (Rolex)

PRECISELY SPEAKING

▲ Rolex's headquarters in Geneva

169

PRECISELY SPEAKING

▲ Fleurier Quality Foundation

Rolex is not the only house to offer its own certification criteria. Patek Philippe opted out of the Geneva Seal to offer its own Patek Philippe Seal. Montblanc has established its 500 Hours Quality Certificate and Jaeger-LeCoultre its 1,000 Hours Control test, to name a few examples. Back on the "public" label side, we also note the Chronofiable tests, which measure the reliability of a watch at the end of an aging simulation process, or the incredibly stringent Fleurier Quality Foundation Label (LQF). Any watch presented to the LQF must first earn a COSC certificate and the Chronofiable stamp before being tested in real-wear conditions, including shocks, for a daily deviation in rate that must not exceed 0 to +5 seconds. That's not even counting the technical, aesthetic and original requirements included in the certification.

PRECISELY SPEAKING

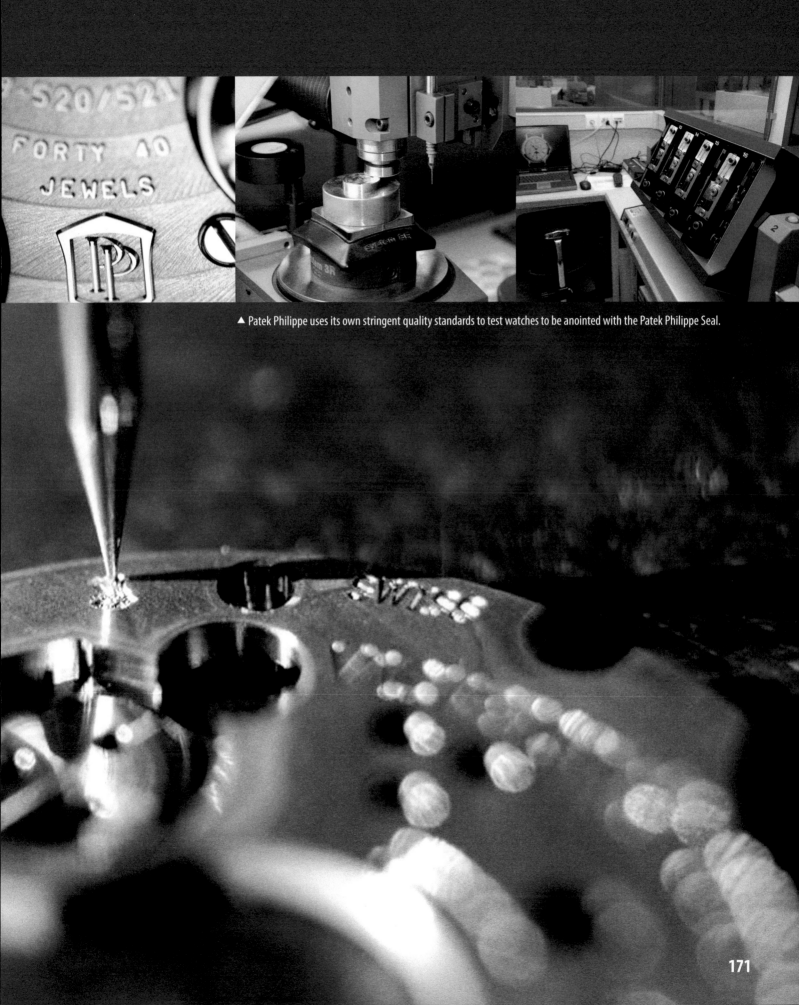

▲ Patek Philippe uses its own stringent quality standards to test watches to be anointed with the Patek Philippe Seal.

CANNES • DUBAI • GENEVA • GSTAAD • KUWAIT • LONDON • NEW YORK
PARIS • PORTO CERVO • ROME • SEOUL • S^T MORITZ

www.degrisogono.com

Ventaglio

PRECISELY SPEAKING

▲ **L.U.C Triple Certification Tourbillon**
(Chopard)

Chopard is one of the most daring watchmakers when it comes to new labels: in 2011, the company unveiled its L.U.C Triple Tourbillon Certification, "the world's first timepiece to simultaneously bear the COSC and Fleurier Quality Foundation certifications as well as the Poinçon de Genève quality hallmark." In terms of traditional mechanics for mass-produced watches, we have probably reached the limits of precision. But as we can easily guess: watchmakers will never be satisfied!

PRECISELY SPEAKING

CANNES • DUBAI • GENEVA • GSTAAD • KUWAIT • LONDON • NEW YORK
PARIS • PORTO CERVO • ROME • SEOUL • ST MORITZ

www.degrisogono.com

Allegra Toi & Moi

A. LANGE & SÖHNE
GLASHÜTTE I/SA

PURE PRECISION

Long renowned for its perfectionist approach to haute horology, A. Lange & Söhne continues to populate its collections with **GORGEOUS, IMPECCABLY PRODUCED FEATS OF WATCHMAKING**.

A high-contrast take on one of the most popular and useful complications, A. Lange & Söhne's 1815 Chronograph takes visual cues from 19th-century pocket watches, with a modern result. The railway-track minute scale and symmetrical subdials emphasize a classic sense of balance, with the dramatic black dial (crafted in solid silver) setting off the white markings, rhodiumed hands, and 39.5mm white-gold case. The chronograph measures times to one-fifth of a second, thanks to the L951.5 movement, which beats at 18,000 vph. Around the circumference of the dial, A. Lange & Söhne has added another useful function: a pulsimeter, which allows the wearer to measure his or her own (or someone else's) heart rate. Fitting the dictates of the calendar watch to the traditional aesthetic of the 1815 collection, A. Lange & Söhne's 1815 Annual Calendar is an instant classic. The manual-winding movement, the new L051.3, powers display of the seconds, minutes, hours, days, dates, months, and moonphase, efficiently enough to dispose of a 72-hour power reserve. The 40mm rose-gold case shelters three recessed pushbuttons to independently adjust the day, month, and moonphase, as well as a pushbutton that advances all indications at once. Though the annual calendar, as its name suggests, must be adjusted once a year on March 1, the beautifully made moonphase indication will be accurate for 122.6 years.

▲ **1815 CHRONOGRAPH**
A. Lange & Söhne combines modern sleekness and pocket-watch classicism in its black and white 1815 Chronograph.

◄ **1815 ANNUAL CALENDAR**
The calendar indications on the 1815 Annual Calendar are efficiently laid out and readable at a glance.

A. LANGE & SÖHNE

The oversize date is a signature aesthetic flourish of A. Lange & Söhne, placing the most important information—time and date—front and center.

Addressing the perpetual calendar complication with its characteristic refinement, A. Lange & Söhne garnered acclaim for its Langematik Perpetual, which exemplifies the brand's perfectionist approach. In contrast to an annual calendar, the perpetual calendar takes leap years into account, meaning that the Langematik Perpetual requires no adjustment until March 1, 2100. The leap-year cycle indication discreetly overlaps the month subdial at 3 o'clock, leading the eye to the highly precise moonphase and small seconds subdial at 6 o'clock and the day display (coupled with a 24-hour time displays and day/night indication) at 9 o'clock. Reigning over the black dial from its position at 12 o'clock, the oversize date is a signature aesthetic flourish of A. Lange & Söhne, placing the most important information—time and date—front and center.

Some consider the quest for ever-thinner movements and cases to be a form of complication in itself, in its striving for the heights of horological mastery. The Saxonia Thin is a real achievement in this arena, measuring just 5.9mm in thickness for the entire case and 2.9mm for the movement alone. Paring away all but the absolutely necessary, the Saxonia Thin displays the hours and minutes within a case diameter of 40mm, and the new L093.1 caliber through the sapphire crystal caseback. The thinnest Lange caliber ever made, the L093.1 meets all the German watchmaker's exacting requirements for performance and decoration, with Glashütte ribbing and on the three-quarter plate, blued screws securing gold chatons, a sunray finish on the wheels, and hand-engraving on the escapement.

▲ **LANGEMATIK PERPETUAL**
The L922.1 SAX-O-MAT movement drives the Langematik Perpetual, housed in a sturdy white-gold case.

▶ **SAXONIA THIN**
The rose-gold case, hands, and indexes of the Saxonia Thin complement its solid silver argenté dial.

A. LANGE & SÖHNE

DATOGRAPH PERPETUAL — REF. 410.038

Movement: manually-wound Lange manufacture L952.1 caliber; 36-hour power reserve; 45 jewels.
Functions: hours, minutes; chronograph: running seconds at 9, 30-minute counter at 3; perpetual calendar: days at 9, month at 3; date at 12; moonphase at 6.
Case: 18K white gold; Ø 41mm, thickness: 13.5mm; sapphire crystal; sapphire crystal caseback.
Dial: silver.
Strap: black hand-stitched alligator leather; white-gold Lange prong buckle.
Price: available upon request.

DATOGRAPH UP/DOWN — REF. 405.035

Movement: manually-wound Lange manufacture L951.6 caliber; 60-hour power reserve; 46 jewels.
Functions: hours, minutes; chronograph: running seconds at 9, 30-minute counter at 3, flyback; power reserve indicator at 6; big date at 12.
Case: 18K white gold; Ø 41mm, thickness: 30.6mm; sapphire crystal; sapphire crystal caseback.
Dial: black.
Strap: black hand-stitched alligator leather.
Price: available upon request.

GRAND LANGE 1 — REF. 117.025

Movement: manually-wound Lange manufacture L095.1 caliber; 72-hour power reserve; 42 jewels; 21,600 vph.
Functions: hours, minutes; small seconds; power reserve indicator; big date.
Case: platinum; Ø 40.9mm, thickness: 8.8mm; sapphire crystal; sapphire crystal caseback.
Dial: silver.
Strap: black hand-stitched alligator leather.
Price: available upon request.

GRANDE LANGE 1 MOON PHASE — REF. 139.032

Movement: manually-wound Lange manufacture caliber L095.3; 72-hour power reserve; 45 jewels.
Functions: hours, minutes; small seconds at 4; power reserve indicator at 3; big date at 2.
Case: pink gold; Ø 41mm, thickness: 9.5mm; sapphire crystal; sapphire crystal caseback.
Dial: silver.
Strap: red-brown hand-stitched alligator leather.
Price: available upon request.

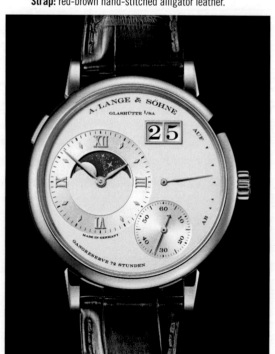

A. LANGE & SÖHNE

GRANDE LANGE 1 MOON PHASE — REF. 139.025

Movement: manually-wound Lange manufacture caliber L095.3; 72-hour power reserve; 45 jewels.
Functions: hours, minutes; small seconds at 4; power reserve indicator at 3; big date at 2.
Case: platinum; Ø 41mm, thickness: 9.5mm; sapphire crystal; sapphire crystal caseback.
Dial: silver.
Strap: black hand-stitched alligator leather.
Price: available upon request.

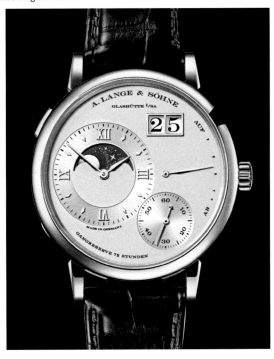

LANGE 1 — REF. 191.032

Movement: manually-wound Lange manufacture L121.1 caliber; 72-hour power reserve; 43 jewels.
Functions: hours, minutes; small seconds; power reserve indicator at 3; big date at 1.
Case: pink gold; Ø 38.5mm, thickness: 9.8mm; sapphire crystal; sapphire crystal caseback.
Dial: silver.
Strap: red-brown hand-stitched alligator leather.
Price: available upon request.

LANGE 1 MOON PHASE — REF. 192.029

Movement: manually-wound Lange manufacture L121.3 caliber; 72-hour power reserve; 47 jewels.
Functions: hours, minutes; small seconds; power reserve indicator at 3; big date at 1.
Case: white gold; Ø 38.5mm, thickness: 10.2mm; sapphire crystal; sapphire crystal caseback.
Dial: silver.
Strap: black hand-stitched alligator leather.
Price: available upon request.

LANGE 1 TIME ZONE — REF. 116.032

Movement: manually-wound Lange manufacture L031.1 caliber.
Functions: hours, minutes; small seconds with stop seconds and patented zero-reset function; perpetual calendar: date, day, month, moonphase and leap year; day/night indicator.
Case: 18K pink gold; Ø 41.9 mm.
Dial: solid silver; gold appliqués.
Strap: red-brown hand-stitched alligator leather; Lange prong buckle.
Price: available upon request.

A. LANGE & SÖHNE

LITTLE LANGE 1 — REF. 181.037

Movement: manually-wound Lange manufacture caliber L121.1; 72-hour power reserve; 43 jewels.
Functions: hours, minutes; small seconds; power reserve indicator at 3; big date at 2.
Case: pink gold; Ø 36.8mm, thickness: 9.5mm; sapphire crystal; sapphire crystal caseback.
Dial: gold.
Strap: taupe alligator leather.
Price: available upon request.

LITTLE LANGE 1 MOON PHASE — REF. 182.030

Movement: manually-wound Lange manufacture L121.2 caliber; 72-hour power reserve.
Functions: hours, minutes; small seconds; power reserve indicator at 3; big date at 2.
Case: pink gold; Ø 36.8mm.
Dial: silver.
Strap: white alligator leather.
Price: available upon request.

RICHARD LANGE TOURBILLON 'POUR LE MÉRITE' — REF. 760.032

Movement: manually-wound Lange manufacture caliber L044.1; Ø 31.6mm, thickness: 6mm; 36-hour power reserve; 33 jewels.
Functions: hours, minutes; small seconds; tourbillon.
Case: platinum; Ø 40.5mm, thickness: 10.7mm; sapphire crystal; sapphire crystal caseback.
Dial: 3-part enamel dial.
Strap: hand-stitched alligator leather; Lange prong buckle.
Note: limited edition of 50 pieces.
Price: available upon request.

SAXONIA ANNUAL CALENDAR — REF. 330.026

Movement: automatic-winding Lange manufacture caliber L085.1 Sax-O-Mat; 46-hour power reserve; 43 jewels.
Functions: hours, minutes; small seconds with patented zero-reset function; annual calendar: big date at 12, month at 3, moonphase at 6, day at 9.
Case: white gold; Ø 38.5mm, thickness: 9.8mm; sapphire crystal; sapphire crystal caseback.
Dial: silver.
Strap: hand-stitched alligator leather.
Price: available upon request.

A. LANGE & SÖHNE

SAXONIA MOON PHASE — REF. 384.032

Movement: manually-wound Lange manufacture L086.5 caliber; 72-hour power reserve; 40 jewels.
Functions: hours, minutes; small seconds; big date at 12; moonphase at 6.
Case: pink gold; Ø 40mm, thickness: 5.2mm; sapphire crystal; sapphire crystal caseback.
Dial: silver.
Strap: red-brown hand-stitched alligator leather; pink-gold prong buckle.
Price: available upon request.

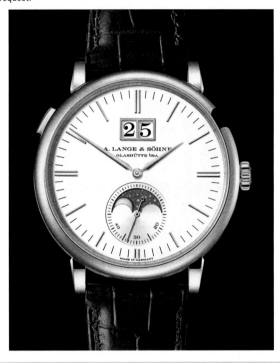

SAXONIA MOON PHASE — REF. 384.031

Movement: manually-wound Lange manufacture L086.5 caliber; 72-hour power reserve; 40 jewels.
Functions: hours, minutes; small seconds; big date at 12; moonphase at 6.
Case: pink gold; Ø 40mm, thickness: 9.8mm; sapphire crystal; sapphire crystal caseback.
Dial: black.
Strap: black hand-stitched alligator leather.
Price: available upon request.

ZEITWERK — REF. 140.032

Movement: manually-wound Lange manufacture L043.1 caliber; 36-hour power reserve; 66 jewels.
Functions: jumping hours at 9; jumping minutes at 3; small seconds at 6; power reserve indicator at 12.
Case: pink gold; Ø 41.9mm, thickness: 12.6mm; sapphire crystal; sapphire crystal caseback.
Dial: silver.
Strap: hand-stitched alligator leather.
Price: available upon request.

ZEITWERK STRIKING TIME — REF. 145.029

Movement: manually-wound Lange manufacture L043.2 caliber; 36-hour power reserve; 78 jewels.
Functions: jumping hours at 9; jumping minutes at 3; small seconds at 6; power reserve indicator at 12; chimes at quarter and full hours.
Case: white gold; Ø 44.2mm, thickness: 13.8mm; sapphire crystal; sapphire crystal caseback.
Dial: black.
Strap: black hand-stitched alligator leather.
Price: available upon request.

AUDEMARS PIGUET
Le Brassus

NEW DEVELOPMENTS ON FIRM FOUNDATIONS

Always **AT THE VANGUARD OF GORGEOUS HAUTE HOROLOGY**, Audemars Piguet presents a bevy of models that push the envelope when it comes to luxury and technical sophistication, while building on the foundations of its iconic Millenary and Royal Oak collections.

The Millenary collection has long been a touchstone of high-end femininity, and two models develop that vision while staying true to its roots. The off-center white opal dial, which boasts no hour markers, is punctuated by a small seconds display at 7:30. At a just-right size of 39.5x35.4mm, the hammered and satin-finished case of the Millenary is available in 18-karat rose or white gold. The most intriguing development is the Polish mesh bracelet, crafted from gold wire "springs" that are woven together, similar to the Milanese style, but with a more complex, richer construction.

The Royal Oak Concept collection launched in 2002, and 16 years later, the collection graced its first ladies' model with another first: Audemars Piguet's first flying tourbillon. The extraordinary architecture of this piece combines haute horology and haute jewelry, glistening with either baguette-cut diamonds or brilliant-cut diamonds. The visual intrigue of the dial is heightened by a stunning tourbillon that seems to float weightless amidst this splendor. The 38.5mm 18- karat white-gold case frames a fascinating look at satin-finished bridges, polished angles and white lacquered decorations through the caseback, while the dial allows a peek at a masterfully done openworked barrel at 11 o'clock, shaped like a snowflake. A hand-stitched "large square scale" white alligator strap with diamond-set 18-karat white-gold AP folding clasp can be swapped out with an additional white "constellation" rubber strap delivered with the watch.

◀ **MILLENARY** *(far left)*
The epitome of feminine luxury, the new Millenary models expertly use a palette of diamonds, rose gold, white gold and mother-of-pearl to express a vision of muted opulence.

◀ **ROYAL OAK CONCEPT FLYING TOURBILLON**
Baguettes or brilliants, the diamonds set on this Royal Oak Concept watch resemble icicles, which join with the white lacquered accents and snowflake-shaped barrel to evoke winter in the Vallée de Joux.

AUDEMARS PIGUET

A futuristic, angular dial configuration highlights the Royal Oak Concept flying tourbillon at 9 o'clock, second time zone at 3 o'clock and a crown HNR display at 6 o'clock.

The men's version of the Royal Oak Concept also boasts a flying tourbillon, as well as the practical second time zone indication. A futuristic, angular dial configuration highlights the tourbillon at 9 o'clock, second time zone at 3 o'clock, and an HNR display (which shows the crown position) at 6 o'clock. The tourbillon movement boasts 237 hours of power reserve and beats at 21,600 vph. Rose-gold-toned accents throughout lend a depth and complexity to the dial design, which is otherwise crafted in sandblasted titanium (the dial) and sandblasted black titanium (the bridges). The sapphire crystal caseback joins the openworked dial to present a full picture of the marriage between cutting-edge technology and a riveting new aesthetic.

Nothing is more chic than black, and the new Royal Oak Tourbillon Chronograph Openworked uses the classic shade throughout. The 44mm black ceramic case and bezel join a blackened mainplate and rhodium-plated barrel, balance wheel, tourbillon bridge and gear train for a forceful visual statement. Possessing a tourbillon, chronograph, hours, minutes and small seconds indications, this model needs nothing else to draw the eye. A hint of color is restricted to the seconds hand, which matches a color highlight on the black rubber strap; this timepiece is also available with a black ceramic bracelet or a two-tone black and gray rubber strap.

▲ **ROYAL OAK CONCEPT FLYING TOURBILLON GMT**
With a flying tourbillon, 24-hour GMT indication and crown position indicator within a 44mm sandblasted titanium case, the Royal Oak Concept – Flying Tourbillon GMT wraps technical sophistication in a fashion-forward aesthetic.

▶ **ROYAL OAK TOURBILLON CHRONOGRAPH OPENWORKED**
The luminescent coating on this watch's Royal Oak hands makes them stand out beautifully against the openworked, satin-brushed black dial.

AUDEMARS PIGUET

MILLENARY FROSTED GOLD   REF. 77244OR.GG.1272OR.01

Movement: manual-winding 5201 Manufacture caliber; 49-hour power reserve; 157 components; 19 jewels; 21,600 vph.
Functions: hours, minutes; small seconds between 7 and 8.
Case: hammered and satin-finished 18K pink gold; Ø 39.5mm, thickness: 9.8mm; crown set with a sapphire cabochon; antireflective sapphire crystal and caseback; water resistant to 2atm.
Dial: white opal disc; pink-gold hands.
Bracelet: 18K pink gold polish mesh; folding clasp.

MILLENARY HAND-WOUND   REF. 77247BC.ZZ.1272BC.01

Movement: manual-winding 5201 Manufacture caliber; 49-hour power reserve; 157 components; 19 jewels; 21,600 vph.
Functions: hours, minutes; small seconds between 7 and 8.
Case: 18K white gold; Ø 39.5mm, thickness: 9.8mm; bezel and lugs set with 116 brilliant-cut diamonds (0.6 carat); crown set with a pink sapphire cabochon; antireflective sapphire crystal and caseback; water resistant to 2atm.
Dial: white mother-of-pearl; anthracite Roman numerals; pink-gold hands.
Bracelet: 18K white gold polish mesh; folding clasp.

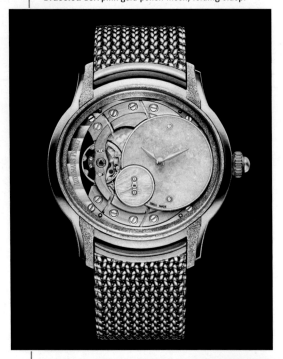

ROYAL OAK DOUBLE BALANCE WHEEL OPENWORKED   REF. 15467OR.OO.1256OR.01

Movement: automatic-winding 3132 Manufacture caliber; 45-hour power reserve; 245 components; 38 jewels; 21,600 vph; gold oscillating weight.
Functions: hours, minutes, seconds.
Case: 18K pink gold; Ø 37mm, thickness: 10mm; screw-down crown; antireflective sapphire crystal and caseback; water resistant to 5atm.
Dial: openworked rhodium-toned; pink-gold applied hour markers; pink-gold Royal Oak hands with luminescent coating.
Bracelet: 18K pink gold; AP folding clasp.

ROYAL OAK TOURBILLON CHRONOGRAPH OPENWORKED   REF. 26343CE.OO.1247CE.01

Movement: automatic-winding 2936 Manufacture caliber; 72-hour power reserve; 299 components; 28 jewels; 21,600 vph.
Functions: hours, minutes; small seconds at 9; chronograph: central seconds hand, 30-minute counter at 3; tourbillon at 6.
Case: black ceramic; Ø 44mm, thickness: 13.2mm; antireflective sapphire crystal and caseback; water resistant to 2atm.
Dial: satin-brushed openworked black; white-gold applied hour markers; white-gold Royal Oak hands with luminescent coating.
Bracelet: black ceramic; titanium AP folding clasp; two additional rubber straps: one in full black rubber and one in two-tone rubber, black on top and gray on the sides and underside.

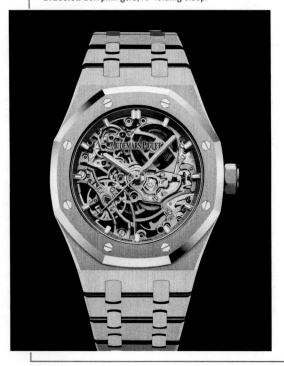

AUDEMARS PIGUET

ROYAL OAK TOURBILLON EXTRA-THIN   REF. 26522ST.00.1220ST.01

Movement: manual-winding 2924 Manufacture caliber; 70-hour power reserve; 216 components; 25 jewels; 21,600 vph.
Functions: hours, minutes; power reserve indicator; tourbillon at 6.
Case: stainless steel; Ø 41mm, thickness: 9mm; antireflective sapphire crystal and caseback; water resistant to 5atm.
Dial: purple engine-turned "Tapisserie Evolutive" pattern; white-gold applied hour markers; white-gold Royal Oak hands with luminescent coating.
Bracelet: stainless steel; AP folding clasp.

ROYAL OAK PERPETUAL CALENDAR   REF. 26584OR.00.1220OR.01

Movement: automatic-winding 5134 Manufacture caliber; 40-hour power reserve; 374 components; 38 jewels; 19,800 vph; gold oscillating weight.
Functions: hours, minutes; perpetual calendar: day at 9, date at 3, month and leap year at 12, week via central hand; moonphase at 6.
Case: 18K pink gold; Ø 41mm, thickness: 9.5mm; screw-down crown; antireflective sapphire crystal and caseback; water resistant to 2atm.
Dial: pink gold-toned engine-turned "Grande Tapisserie" pattern; blue counters; pink-gold applied hour markers; pink-gold Royal Oak hands with luminescent coating.
Bracelet: 18K pink gold; AP folding clasp.
Note: limited edition of 100 pieces.

ROYAL OAK FROSTED GOLD SELFWINDING   REF. 15454BC.GG.1259BC.03

Movement: automatic-winding 3120 Manufacture caliber; 60-hour power reserve; 280 components; 40 jewels; 21,600 vph; 22K gold oscillating weight.
Functions: hours, minutes, seconds; date at 3.
Case: 18K white gold; Ø 37mm, thickness: 9.8mm; screw-down crown; antireflective sapphire crystal and caseback; water resistant to 5atm.
Dial: black engine-turned "Grande Tapisserie" pattern; white-gold applied hour markers; Royal Oak hands with luminescent coating.
Bracelet: 18K white gold; AP folding clasp.

ROYAL OAK FROSTED GOLD QUARTZ   REF. 67653OR.GG.1263OR.02

Movement: quartz 2713 caliber; 7 jewels.
Functions: hours, minutes; date at 3.
Case: 18K pink gold; Ø 33mm, thickness: 7mm; antireflective sapphire crystal; water resistant to 5atm.
Dial: pink gold-toned engine-turned "Grande Tapisserie" pattern; pink-gold applied hour markers; Royal Oak hands with luminescent coating.
Bracelet: 18K pink gold; AP folding clasp.

AUDEMARS PIGUET

ROYAL OAK "JUMBO" EXTRA THIN — REF. 15202IP.00.1240IP.01

Movement: automatic-winding 2121 Manufacture caliber; 40-hour power reserve; 247 components; 36 jewels; 19,800 vph; 21K gold oscillating weight.
Functions: hours, minutes; date at 3.
Case: titanium; Ø 39mm, thickness: 8.1mm; polished 950 platinum bezel; antireflective sapphire crystal and caseback; water resistant to 5atm.
Dial: smoked blue engine-turned "Petite Tapisserie" pattern; white-gold applied hour markers; Royak Oak hands with luminescent coating.
Bracelet: titanium; polished 950 platinum links; titanium AP folding clasp.
Note: limited edition of 250 pieces.

ROYAL OAK "JUMBO" EXTRA-THIN — REF. 15202BC.ZZ.1241BC.01

Movement: automatic-winding 2121 Manufacture caliber; 40-hour power reserve; 247 components; 36 jewels; 19,800 vph; 21K gold oscillating weight.
Functions: hours, minutes; date at 3.
Case: diamond-set 18K white gold; Ø 39mm, thickness: 8.1mm; antireflective sapphire crystal and caseback; water resistant to 5atm.
Dial: 18K white gold; set with 384 brilliant-cut diamonds (2.44 carats); white-gold applied hour markers; Royal Oak hands with luminescent coating.
Bracelet: diamond-set 18K white gold; AP folding clasp.
Note: case and bracelet set with 1,102 brilliant-cut diamonds (7.09 carats).

ROYAL OAK CONCEPT FLYING TOURBILLON — REF. 26227BC.ZZ.D011CR.01

Movement: manual-winding 2951 Manufacture caliber; 72-hour power reserve; 255 components; 17 jewels; 21,600 vph; decoration: set with 9 brilliant-cut diamonds (0.02 carat).
Functions: hours, minutes; tourbillon at 6.
Case: 18K white gold; Ø 38.5mm, thickness: 11.4mm; case set with brilliant-cut diamonds; white-gold inner bezel set with brilliant-cut diamonds; crown set with a translucent sapphire cabochon; antireflective sapphire crystal and caseback; water resistant to 2atm.
Dial: openworked 18K white gold; set with 62 brilliant-cut diamonds (0.16 carat); openworked white-gold hands with luminescent coating.
Strap: hand-stitched white alligator leather; 18K white-gold AP folding clasp set with brilliant-cut diamonds.
Note: case and buckle set with 397 brilliant-cut diamonds (3.49 carats).

ROYAL OAK CONCEPT FLYING TOURBILLON GMT — REF. 26589IO.00.D002CA.01

Movement: manual-winding 2954 Manufacture caliber; 237-hour power reserve; 348 components; 24 jewels; 21,600 vph.
Functions: hours, minutes; GMT at 3; tourbillon at 9.
Case: sandblasted titanium; Ø 44mm, thickness: 16.1mm; black ceramic bezel; screw-down crown and pushpiece; antireflective sapphire crystal and caseback; water resistant to 10atm.
Dial: openworked; second time zone on sapphire plate at 3; crown position indicator at 6; white-gold Royal Oak hands with luminescent coating.
Strap: black rubber; sandblasted titanium AP folding clasp.

AUDEMARS PIGUET

ROYAL OAK OFFSHORE SELFWINDING CHRONOGRAPH REF. 26470ST.OO.A099CR.01

Movement: automatic-winding 3126/3840 Manufacture caliber; 50-hour power reserve; 365 components; 59 jewels; 21,600 vph; 22K gold oscillating weight.
Functions: hours, minutes; small seconds at 12; date at 3; chronograph: central seconds hand, 12-hour counter at 6, 30-minute counter at 9.
Case: stainless steel; Ø 42mm, thickness: 14.5mm; blue inner bezel; black ceramic pushpieces and screw-down crown; antireflective sapphire crystal and caseback; water resistant to 10atm.
Dial: brown with "Méga Tapisserie" pattern; blue counters; Arabic numerals with luminescent coating; white-gold Royal Oak hands with luminescent coating.
Strap: hand-stitched brown alligator leather; stainless steel pin buckle.

ROYAL OAK OFFSHORE SELFWINDING CHRONOGRAPH REF. 26231ST.ZZ.D010CA.01

Movement: automatic-winding 2385 caliber; 40-hour power reserve; 304 components; 37 jewels; 21,600 vph; 18K gold oscillating weight.
Functions: hours, minutes; small seconds at 6; date between 4 and 5; chronograph: central seconds hand, 30-minute counter at 3, 12-hour counter at 9.
Case: stainless steel; Ø 37mm, thickness: 12.4mm; bezel set with 32 brilliant-cut diamonds (1.02 carats); screw-down crown; antireflective sapphire crystal; water resistant to 5atm.
Dial: silver-toned with "Lady Tapisserie" pattern; blue counters; white-gold applied hour markers; Royal Oak hands with luminescent coating.
Strap: white rubber; stainless steel AP folding clasp.

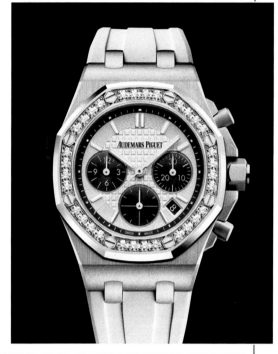

ROYAL OAK OFFSHORE TOURBILLON CHRONOGRAPH REF. 26421OR.OO.A002CA.01

Movement: manual-winding 2947 Manufacture caliber; 173-hour power reserve; 353 components; 30 jewels; 21,600 vph.
Functions: hours, minutes; chronograph: central seconds hand, 30-minute counter; tourbillon at 9.
Case: 18K pink gold; Ø 45mm, thickness: 16.1mm; black ceramic screw-down crown and pushpieces; antireflective sapphire crystal and caseback; water resistant to 10atm.
Dial: black; pink-gold hands with luminescent coating.
Strap: black rubber; 18K pink-gold pin buckle.
Note: limited edition of 50 pieces.

ROYAL OAK OFFSHORE DIVER REF. 15710ST.OO.A032CA.01

Movement: automatic-winding 3120 Manufacture caliber; 60-hour power reserve; 280 components; 40 jewels; 21,600 vph; 22K gold oscillating weight.
Functions: hours, minutes, seconds; date at 3.
Case: stainless steel; Ø 42mm, thickness: 14.1mm; turquoise blue rotating inner bezel; turquoise blue rubber-clad screw-down crowns; antireflective sapphire crystal and caseback; water resistant to 30atm.
Dial: turquoise blue with "Méga Tapisserie" pattern; white-gold applied hour markers; Royal Oak hands with luminescent coating.
Strap: turquoise blue rubber; stainless steel pin buckle.

NEW LIVES FOR A LEGEND

A pioneer in the world of diving watches, Blancpain introduced the Fifty Fathoms collection in 1953, and the Bathyscaphe model in 1956. Three new models in the storied line break **NEW GROUND TECHNICALLY, WHILE HONORING ITS HISTORY**.

▲ **FIFTY FATHOMS GRANDE DATE**
Controlled via the crown, the large double-window date function provides a new feature to the Fifty Fathoms collection.

The Fifty Fathoms Grande Date brings a new element to the iconic Fifty Fathoms collection, moving the date from 4 o'clock to 6 o'clock on the dial, increasing its size, and framing it in a double window. Already a popular feature in the Villeret collection, the large date display transforms the dial of the Fifty Fathoms Grande Date. The greater size of the numerals enhances readability, and the signature mechanism within changes the date instantly at midnight, without drawing excessive energy from the watch's running train. In keeping with Blancpain's policy of using movements constructed completely in-house, the base caliber 1315 has been entirely created by the Swiss manufacture, and boasts three barrels for a generous five-day power reserve. A hairspring crafted in silicon, a non-magnetic material, obviates the need for a soft iron inner case, allowing the watch to benefit from a transparent sapphire crystal caseback. Fifty Fathoms' characteristic unidirectional rotating bezel lends a gentle sense of roundness to the case, with the scratch-resistant protection of the sapphire used in its construction proving that sleek and sturdy can coexist, and the 45mm satin-finished titanium case combines reliability and light weight. The rotating bezel is also an expression of the Fifty Fathoms' diving identity, offering as it does a way to see dive time at a glance.

BLANCPAIN

Vintage visual touches guide the new Bathyscaphe Day Date 70s, a retro-styled yet modern limited edition watch.

The Fifty Fathoms collection and its original Bathyscaphe model launched in the 1950s, but its style has always adapted to the era in which it finds itself. In the 1970s, for example, came a distinctive look, with Arabic numerals at five-minute intervals, rectangular indexes, windows for the day and date at 3 o'clock, and the period's embrace of earth tones. These visual touches guide the new Bathyscaphe Day Date 70s, a retro-styled but nonetheless modern watch that is available in a limited edition of 500 pieces. The smoky ombré dial makes the face appear larger, while enhancing readability. Though the dial expresses a 20th-century style, the watch draws upon Blancpain's technical achievements well into the 21st. The ceramic insert on the unidirectional rotating bezel sports indexes filled with Liquidmetal®, and the 281-component 1315DD movement that beats inside is up to the latest standards, providing a power reserve of 120 hours. The 43mm satin-brushed steel case boasts water resistance to 300m as well as a sapphire crystal caseback.

The utilitarian diving watch joins forces with the world of complications on the Fifty Fathoms Bathyscaphe Quantième Annuel, which places an annual calendar in the iconic collection. A 43mm satin-finished stainless steel case protects the new Blancpain caliber inside, the 6054.P, to a depth of 300m. Intuitively arranged apertures on the dial reveal indications for day, date, and month. As an annual calendar, the Fifty Fathoms Bathyscaphe Quantième Annuel accounts for the varying lengths of 11 months, requiring a manual adjustment from the wearer just once per year at the end of February. Staying true to its diving watch legacy, this annual calendar model features a unidirectional rotating bezel in black ceramic and Liquidmetal®, subtly contrasting with the meteor gray dial. Combining everyday functionality and diving robustness, the Fifty Fathoms Bathyscaphe Quantième Annuel is available on a variety of straps.

▲ **FIFTY FATHOMS DAY DATE 70s**
The vintage-styled Fifty Fathoms Day Date 70s is housed in a 43mm satin-finished steel case, water resistant to 300m.

▶ **FIFTY FATHOMS BATHYSCAPHE**
The first Bathyscaphe model to feature an annual calendar, the Fifty Fathoms Bathyscaphe Quantième Annuel adds a touch of versatility to the iconic diving watch collection.

BLANCPAIN

FIFTY FATHOMS GRANDE DATE — REF. 5050-12B30-B52A

Movement: automatic-winding; Ø 32mm, thickness: 7.15mm; 120-hour power reserve; 262 components; 44 jewels.
Functions: hours, minutes, seconds; date at 6.
Case: satin-brushed titanium; Ø 45mm, thickness: 16.27mm; unidirectional sapphire crystal bezel; sapphire crystal caseback; water resistant to 30atm.
Dial: black.
Strap: black sail-canvas; pin buckle.
Suggested price: $17,500

FIFTY FATHOMS BATHYSCAPHE DAY DATE 70S — REF. 5052-1110-063A

Movement: automatic-winding; Ø 34.75mm, thickness: 6.6mm; 120-hour power reserve; 281 components; 37 jewels.
Functions: hours, minutes, seconds; day and date at 3.
Case: satin-brushed steel; Ø 43mm, thickness: 14.25mm; unidirectional steel bezel with ceramic insert; sapphire crystal caseback; water resistant to 30atm.
Dial: graduated gray coloring; Liquidmetal® hour markers.
Strap: brown barrenia lined with rubber; pin buckle.
Suggested price: $12,700

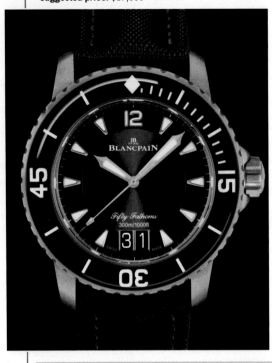

FIFTY FATHOMS BATHYSCAPHE ANNUAL CALENDAR — REF. 5071-1110-B52A

Movement: automatic-winding; Ø 32mm, thickness: 5.73mm; 72-hour power reserve; 362 components; 34 jewels.
Functions: hours, minutes, seconds; annual calendar: date at 3, day at 2, month at 4.
Case: satin-brushed steel; Ø 43mm, thickness: 13.46mm; unidirectional steel bezel with ceramic insert; sapphire crystal caseback; water resistant to 30atm.
Dial: meteor gray; Liquidmetal® hour markers.
Strap: black sail-canvas; pin buckle.
Suggested price: $26,100

FIFTY FATHOMS BATHYSCAPHE ANNUAL CALENDAR — REF. 5071-1110-070B

Movement: automatic-winding; Ø 32mm, thickness: 5.73mm; 72-hour power reserve; 362 components; 34 jewels.
Functions: hours, minutes, seconds; annual calendar: date at 3, day at 2, month at 4.
Case: satin-brushed steel; Ø 43mm, thickness: 13.46mm; unidirectional steel bezel with ceramic insert; sapphire crystal caseback; water resistant to 3atm.
Dial: meteor gray; Liquidmetal® hour markers.
Bracelet: stainless steel; folding buckle.
Suggested price: $28,800

BLANCPAIN

FIFTY FATHOMS BATHYSCAPHE COMPLETE CALENDAR MOON PHASE REF. 5054-1110-070B

Movement: automatic-winding; Ø 32mm, thickness: 5.48mm; 72-hour power reserve; 321 components; 28 jewels.
Functions: hours, minutes, seconds; moonphase at 6; complete calendar: date via central hand, day and month at 12.
Case: satin-brushed steel; Ø 43mm, thickness: 13.9mm; unidirectional steel bezel with ceramic insert; water resistant to 3oatm.
Dial: meteor gray; Liquidmetal® hour markers.
Strap: stainless steel; folding clasp.
Suggested price: $17,400

FIFTY FATHOMS BATHYSCAPHE COMPLETE CALENDAR MOON PHASE REF. 5054-1110-NABA

Movement: automatic-winding; Ø 32mm, thickness: 5.48mm; 72-hour power reserve; 321 components; 28 jewels.
Functions: hours, minutes, seconds; moonphase at 6; complete calendar: date via central hand, day and month at 12.
Case: satin-brushed steel; Ø 43mm, thickness: 13.9mm; unidirectional steel bezel with ceramic insert; water resistant to 30atm.
Dial: meteor gray; Liquidmetal® hour markers.
Strap: black NATO; pin buckle.
Suggested price: $14,800

VILLERET LARGE DATE WITH RETROGRADE DAY OF THE WEEK   REF. 6668-1127-MMB

Movement: automatic-winding; Ø 32mm, thickness: 5.27mm; 72-hour power reserve; 344 components; 40 jewels.
Functions: hours, minutes, seconds; date at 5; retrograde day at 8.
Case: stainless steel; Ø 40mm, thickness: 11.1mm; sapphire crystal caseback; water resistant to 3atm.
Dial: white.
Bracelet: stainless steel.
Suggested price: $16,100

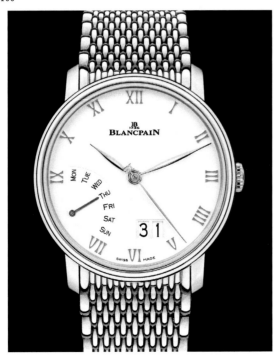

VILLERET LARGE DATE WITH RETROGRADE DAY OF THE WEEK   REF. 6668-3642-55B

Movement: automatic-winding; Ø 32mm, thickness: 5.27mm; 72-hour power reserve; 344 components; 40 jewels.
Functions: hours, minutes, seconds; date at 5; retrograde day at 8.
Case: 18K red gold; Ø 40mm, thickness: 11.1mm; sapphire crystal caseback; water resistant to 3atm.
Dial: opaline.
Strap: brown alligator leather.
Suggested price: $22,900

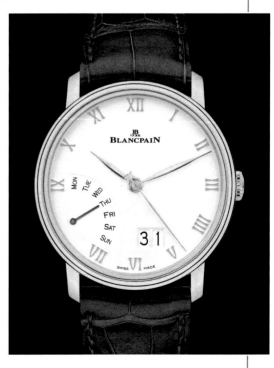

BLANCPAIN

VILLERET FLYING TOURBILLON — REF. 66260-3433-55B

Movement: manual-winding 260MR caliber; Ø 32mm, thickness: 5.85mm; 144-hour power reserve; 263 components; 39 jewels.
Functions: jumping hours; retrograde minutes; tourbillon at 12.
Case: platinum; Ø 42mm, thickness: 11mm; sapphire crystal caseback; water resistant to 3atm.
Dial: white Grand Feu enamel.
Strap: blue alligator leather.
Note: limited edition of 20 pieces.

Suggested price: $180,900

VILLERET FLYING TOURBILLON — REF. 66260-3633-MMB

Movement: manual-winding 260MR caliber; Ø 32mm, thickness: 5.85mm; 144-hour power reserve; 263 components; 39 jewels.
Functions: jumping hours; retrograde minutes; tourbillon at 12.
Case: 18K red gold; Ø 42mm, thickness: 11mm; sapphire crystal caseback; water resistant to 3atm.
Dial: white Grand Feu enamel.
Bracelet: 18K red gold.
Suggested price: $167,400

VILLERET COMPLETE CALENDAR MOON PHASE GMT — REF. 6676-3642-MMB

Movement: automatic-winding; Ø 27mm, thickness: 6mm; 72-hour power reserve; 286 components; 28 jewels.
Functions: hours, minutes; moonphase at 6; complete calendar: date via central hand, month and day at 12; GMT.
Case: 18K red gold; Ø 40mm, thickness: 11.8mm; sapphire crystal caseback; water resistant to 3atm.
Dial: opaline.
Bracelet: 18K red gold.

Suggested price: $45,800

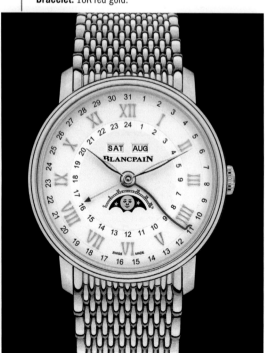

VILLERET COMPLETE CALENDAR MOON PHASE GMT — REF. 6676-1127-55B

Movement: automatic-winding; Ø 27mm, thickness: 6mm; 72-hour power reserve; 286 components; 28 jewels.
Functions: hours, minutes; moonphase at 6; complete calendar: date via central hand, month and day at 12; GMT.
Case: stainless steel; Ø 40mm, thickness: 11.8mm; sapphire crystal caseback.
Dial: white.
Strap: black alligator leather.
Suggested price: $15,900

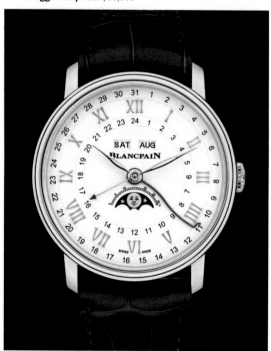

BLANCPAIN

VILLERET PERPETUAL CALENDAR MOON PHASE REF. 6656-1127-55B

Movement: automatic-winding; secured movement; Ø 32mm, thickness: 4.97mm; 72-hour power reserve; 351 components; 32 jewels.
Functions: hours, minutes, seconds; moonphase at 6; perpetual calendar: date at 3, day at 9, month and leap year at 12.
Case: stainless steel; Ø 40mm, thickness: 11.1mm; sapphire crystal caseback; water resistant to 3atm.
Dial: white.
Strap: black alligator leather.
Suggested price: $34,100

VILLERET PERPETUAL CALENDAR MOON PHASE REF. 6656-3642-55B

Movement: automatic-winding; secured movement; Ø 32mm, thickness: 4.97mm; 72-hour power reserve; 351 components; 32 jewels.
Functions: hours, minutes, seconds; moonphase at 6; perpetual calendar: date at 3, day at 9, month and leap year at 12.
Case: 18K red gold; Ø 40mm, thickness: 11.1mm; sapphire crystal caseback; water resistant to 3atm.
Dial: opaline.
Strap: brown alligator leather.
Suggested price: $45,000

WOMENS VILLERET MOON PHASE CALENDAR REF. 6126-4628-MMB

Movement: automatic-winding; secured movement; Ø 23.7mm, thickness: 4.65mm; 72-hour power reserve; 246 components; 20 jewels.
Functions: hours, minutes, seconds; date via central hand; moonphase at 6.
Case: steel; Ø 33.2mm, thickness: 10.2mm; bezel set with diamonds (0.98 carat); sapphire crystal caseback; water resistant to 3atm.
Dial: white; set with 8 diamond hour markers.
Bracelet: stainless steel.
Suggested price: $20,200

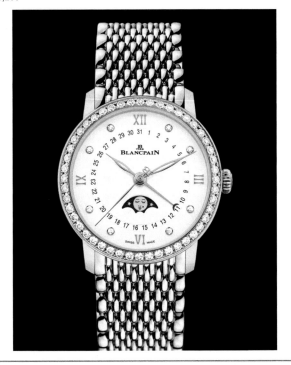

WOMENS VILLERET MOON PHASE CALENDAR REF. 6126-2987-MMB

Movement: automatic-winding; secured movement; Ø 23.7mm, thickness: 4.65mm; 72-hour power reserve; 246 components; 20 jewels.
Functions: hours, minutes, seconds; date via central hand; moonphase at 6.
Case: 18K red gold; Ø 33.2mm, thickness: 10.2mm; bezel set with diamonds (0.98 carat); sapphire crystal caseback; water resistant to 3atm.
Dial: opaline; set with 8 diamond hour markers.
Bracelet: 18K red gold.
Suggested price: $22,900

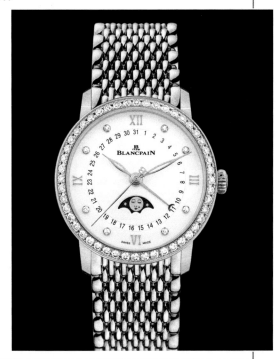

SMOOTH SAILING

The Marine line evolved into a new aesthetic in 2017, maintaining its iconic design while updating key elements for a new age. Three new models **EXPLORE THE POSSIBILITIES AND BOUNDARIES OF THIS REVAMPED STYLE**.

A deceptively simple three-hand model, the Marine 5517 incorporates the new stylistic touches throughout. A date display at 3 o'clock complements the sunburst or wave-engraved dial in understated sophistication, while the Roman numeral hour markers, minute dots, and modern Breguet hands boast a luminescent coating for excellent readability. The 40mm case, crafted in titanium, white gold, or rose gold, sports protective concentric fluting and water resistance to 100m. An open caseback shows off the thematically constructed Cal. 777A movement, which evokes boat docks through its engine-turned decoration and rudder-inspired oscillating weight. Beating at 28,800 vph, the self-winding movement, numbered and signed Breguet, provides a 55-hour power reserve. The Marine 5517 comes in several variations, suiting each taste: a titanium case with sunburst slate gray dial in gold, white-gold case with engraved blue dial in gold, and rose-gold case with engraved, silvered gold dial. To further customize the collection, each model is available with a leather or a rubber strap.

◀ **MARINE 5517**
Available in three case and dial configurations, each with a choice of straps, the Marine 5517 allows the maritime enthusiast to wear his passion with understated flair.

The Marine Chronograph's dial architecture incorporates a twist on classic chronograph layout, with subdials of varying sizes for an intuitive display.

The updated look of the Marine collection extends to the line's chronograph models as well. Developed and enhanced by Abraham-Louis Breguet himself, the guiding, fertile force of the chronograph is given its due recognition in Breguet's Marine Chronograph 5527. The dial architecture incorporates a subtle, yet noticeable, twist on classic chronograph layout, with subdials of varying sizes that make reading the display even more intuitive. Available in the same material configurations as the Marine 5517, the Marine Chronograph 5527 adds the functional and stylistic elements of a large 30-minute counter at 3 o'clock, an underlapping, slightly smaller 12-hour counter at 6 o'clock, and a discrete small seconds display at 9 o'clock. The dial design physically emphasizes the relation between the two chronograph subdials by connecting them, as the chronograph seconds tick by on a central pointer, whose flag design is inspired by maritime codes. An off-kilter date display completes the picture at 4:30. Within the 42.3mm case in white gold, rose gold or titanium, the self-winding movement Cal. 582QA beats at 28,800 vph for a 48-hour power reserve.

With three functions in addition to the time display, Breguet's Marine Alarme Musicale 5547 is a nautically-inclined powerhouse. Its namesake complication is well suited to an active lifestyle with any variety of commitments, providing anything from early wake-up signals to reminders of important appointments. A sizable subdial at 3 o'clock alerts the wearer to the impending reminder, and as the alarm sounds, a ship's bell appears at the top of the dial. The striking mechanism possesses its own power reserve, which appears discreetly between 9 and 12 o'clock, with an arrow indicating the all-red portion of the display when fully wound. Befitting its functionality, the watch face and hands feature luminescent markings for instant legibility, day or night. A second time zone display at 9 o'clock captures "home" time on a 24-hour scale, while the date at 6 o'clock rounds out the fuller picture. The self-winding Cal. 519F/1 movement provides a 45-hour power reserve to the main display and beats at 28,800 vph. Matching the other new models in the Marine collection, the Alarme Musicale sports three case and dial combinations.

▲ **MARINE CHRONOGRAPH 5527**
An extension of Abraham-Louis Breguet's affection for the chronograph, the Marine Chronograph 5527 uses a subtly inventive dial design to underscore the function of each element.

▶ **MARINE ALARME MUSICALE 5547**
Breguet deftly combines an alarm, a second time zone display, date, and time within a fluted 40mm case.

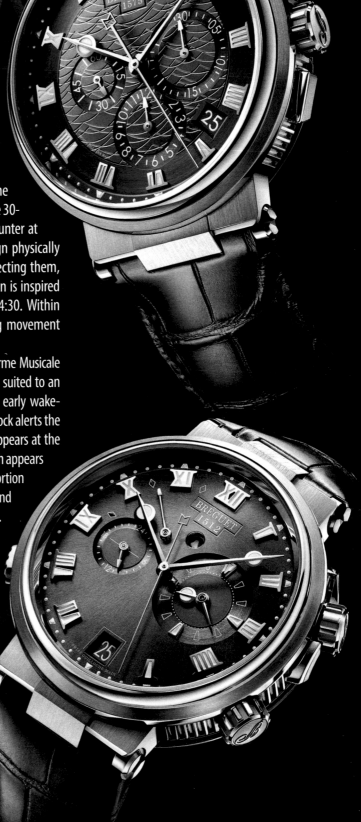

BREGUET

CLASSIQUE TOURBILLON 5367 — REF. 5367BR/29/9WU

Movement: automatic-winding 581 caliber; 16 lines; 80-hour power reserve; 33 jewels; silicon balance spring.
Functions: hours, minutes; small seconds on tourbillon at 5.
Case: rose gold; Ø 42mm; delicately fluted caseband; sapphire crystal caseback; water resistant to 3atm.
Dial: Grand Feu enamel; chapter ring with Breguet Arabic numerals; blued steel Breguet open-tipped hands; signed Breguet.
Strap: alligator leather; gold folding clasp.

Suggested price: $147,500
Also available: platinum case, $161,800 (ref. 5367PT/29/9WU)

CLASSIQUE EXTRA-THIN 5157 — REF. 5157BR/11/9V6

Movement: automatic-winding 502.3 caliber; 12 lines; 45-hour power reserve; 35 jewels; silicon balance spring.
Functions: hours, minutes.
Case: 18K rose gold; Ø 38mm, thickness: 5.45mm; delicately fluted caseband; sapphire crystal caseback; water resistant to 3atm.
Dial: silvered gold; hours chapter with Roman numerals; blued steel Breguet open-tipped hands; individually numbered and signed Breguet.
Strap: alligator leather; pin buckle.

Suggested price: $18,800

REINE DE NAPLES 8908 — REF. 8908BR/5T/864 D00D

Movement: automatic-winding 537 DRL2 caliber; 8 ¾ lines; 45-hour power reserve; 28 jewels; Breguet balance wheel; silicon balance spring.
Functions: hours, minutes; small seconds at 7; moonphase and power reserve indicator at 12.
Case: 18K rose gold; 36.5x28.45mm; delicately fluted caseband; bezel and dial flange set with 128 brilliant-cut diamonds (~0.77 carat); crown set with a cabochon-cut ruby (~0.27 carat); sapphire crystal caseback; water resistant to 3atm.
Dial: Tahitian mother-of-pearl and silvered gold; hours chapter with Roman numerals; blued steel Breguet open-tipped hands; individually numbered and signed Breguet.
Strap: black satin; diamond-set folding clasp.
Suggested price: $36,100
Also available: rose-gold bracelet (ref. 8908BR/5T/J20 D000).

MARINE 5517 — REF. 5517TI/G2/5ZU

Movement: automatic-winding 777A caliber; 15 lines; 55-hour power reserve; 26 jewels; silicon balance spring; numbered and signed Breguet.
Functions: hours, minutes, seconds; date at 3.
Case: titanium; Ø 40mm; fluted caseband; sapphire crystal caseback; water resistant to 10atm.
Dial: sunburst slate gray gold; hours chapter with Roman numerals and luminescent dots; gold facetted Breguet hands with luminescent material; individually numbered and signed Breguet.
Strap: rubber.
Suggested price: $18,500
Also available: leather strap (ref. 5517TI/G2/9ZU).

BREGUET

MARINE 5517 — REF. 5517BB/Y2/9ZU

Movement: automatic-winding 777A caliber; 15 lines; 55-hour power reserve; 26 jewels; silicon balance spring; numbered and signed Breguet.
Functions: hours, minutes, seconds; date at 3.
Case: white gold; Ø 40mm; fluted caseband; sapphire crystal caseback; water resistant to 10atm.
Dial: blue gold; hours chapter with Roman numerals and luminescent dots; gold facetted Breguet hands with luminescent material; individually numbered and signed Breguet.
Strap: leather.
Suggested price: $28,700

MARINE 5517 — REF. 5517BR/12/9ZU

Movement: automatic-winding 777A caliber; 15 lines; 55-hour power reserve; 26 jewels; silicon balance spring; numbered and signed Breguet.
Functions: hours, minutes, seconds; date at 3.
Case: rose gold; Ø 40mm; fluted caseband; sapphire crystal caseback; water resistant to 10atm.
Dial: silvered gold; hours chapter with Roman numerals and luminescent dots; gold facetted Breguet hands with luminescent material; individually numbered and signed Breguet.
Strap: leather.
Suggested price: $28,700

MARINE CHRONOGRAPHE 5527 — REF. 5527BB/Y2/9WV

Movement: automatic-winding 582QA caliber; 14 ½ lines; 48-hour power reserve; 28 jewels; silicon balance spring.
Functions: hours, minutes; small seconds at 9; date between 4 and 5; chronograph: 12-hour counter at 6, 30-minute counter at 3, central seconds hand.
Case: titanium; Ø 42.3mm; fluted caseband; sapphire crystal caseback; water resistant to 10atm.
Dial: sunburst gray gold; hours chapter with Roman numerals and luminescent dots; gold facetted Breguet hands with luminescent material.
Strap: leather; folding clasp.
Suggested price: $22,600
Also available: rubber strap (ref. 5527TI/G2/5WV).

MARINE CHRONOGRAPHE 5527 — REF. 5527/BR/12/5WV

Movement: automatic-winding 582QA caliber; 14 ½ lines; 48-hour power reserve; 28 jewels; silicon balance spring.
Functions: hours, minutes; small seconds at 9; date between 4 and 5; chronograph: 12-hour counter at 6, 30-minute counter at 3, central seconds hand.
Case: rose gold; Ø 42.3mm; fluted caseband; sapphire crystal caseback; water resistant to 10atm.
Dial: silvered gold; hours chapter with Roman numerals and luminescent dots; gold facetted Breguet hands with luminescent material.
Strap: rubber; folding clasp.
Suggested price: $28,700

BREGUET

MARINE CHRONOGRAPHE 5527 — REF. 5527BB/Y25WV

Movement: automatic-winding 582QA caliber; 14 ½ lines; 48-hour power reserve; 28 jewels; silicon balance spring.
Functions: hours, minutes; small seconds at 9; date between 4 and 5; chronograph: 12-hour counter at 6, 30-minute counter at 3, central seconds hand.
Case: white gold; Ø 42.3mm; fluted caseband; sapphire crystal caseback; water resistant to 10atm.
Dial: blue gold; hours chapter with Roman numerals and luminescent dots; gold facetted Breguet hands with luminescent material.
Strap: rubber; folding clasp.
Suggested price: $28,700

MARINE ALARME MUSICALE 5547 — REF. 5547TI/G2/9ZU

Movement: automatic-winding 519F/1 caliber; 12 lines; 45-hour power reserve; 36 jewels; silicon balance spring.
Functions: hours, minutes, seconds; date at 6; second time zone at 9; alarm at 3; alarm activation indicator at 12; power reserve indicator between 9 and 12.
Case: titanium; Ø 40mm; fluted caseband; sapphire crystal caseback; water resistant to 5atm.
Dial: sunburst slate gray gold; hours chapter with Roman numerals and luminescent dots; facetted Breguet hands with luminescent material; numbered and signed Breguet.
Strap: alligator leather; folding clasp.
Suggested price: $28,600
Also available: rubber strap (ref. 5547TI/G2/5ZU).

MARINE ALARME MUSICALE 5547 — REF. 5547BR/12/9ZU

Movement: automatic-winding 519F/1 caliber; 12 lines; 45-hour power reserve; 36 jewels; silicon balance spring.
Functions: hours, minutes, seconds; date at 6; second time zone at 9; alarm at 3; alarm activation indicator at 12; power reserve indicator between 9 and 12.
Case: rose gold; Ø 40mm; fluted caseband; sapphire crystal caseback; water resistant to 5atm.
Dial: silvered gold; hours chapter with Roman numerals and luminescent dots; facetted Breguet hands with luminescent material; numbered and signed Breguet.
Strap: alligator leather; folding clasp.
Suggested price: $40,900

MARINE ALARME MUSICALE 5547 — REF. 5547BB/Y2/9ZU

Movement: automatic-winding 519F/1 caliber; 12 lines; 45-hour power reserve; 36 jewels; silicon balance spring.
Functions: hours, minutes, seconds; date at 6; second time zone at 9; alarm at 3; alarm activation indicator at 12; power reserve indicator between 9 and 12.
Case: white gold; Ø 40mm; fluted caseband; sapphire crystal caseback; water resistant to 5atm.
Dial: blue gold; hours chapter with Roman numerals and luminescent dots; facetted Breguet hands with luminescent material; numbered and signed Breguet.
Strap: alligator leather; folding clasp.
Suggested price: $40,900

BREGUET

TRADITION DAME 7038 — REF. 7038BR/18/9V6 D00D

Movement: automatic-winding 505SR caliber; 14 ½ lines; 50-hour power reserve; 38 jewels; Breguet silicon balance spring.
Functions: hours, minutes; power reserve indicator at 10.
Case: 18K rose gold; Ø 37mm; bezel set with 68 brilliant-cut diamonds (~0.895 carat); crown set with a watch movement jewel; delicately fluted caseband; sapphire crystal caseback; water resistant to 3atm.
Dial: natural white mother-of-pearl; hours chapter with Arabic Breguet numerals; gold Breguet open-tipped hands; individually numbered and signed Breguet.
Strap: alligator leather; gold pin buckle set with 19 brilliant-cut diamonds (~0.135 carat).
Suggested price: $38,100

TRADITION DAME 7038 — REF. 7038BB/1T/9V6 D00D

Movement: automatic-winding 505SR caliber; 14 ½ lines; 50-hour power reserve; 38 jewels; Breguet silicon balance spring.
Functions: hours, minutes; power reserve indicator at 10.
Case: 18K white gold; Ø 37mm; bezel set with 68 brilliant-cut diamonds (~0.895 carat); crown set with a watch movement jewel; delicately fluted caseband; sapphire crystal caseback; water resistant to 3atm.
Dial: Tahitian mother-of-pearl; hours chapter with Roman numerals; blued steel Breguet open-tipped hands; individually numbered and signed Breguet.
Strap: alligator leather.
Suggested price: $38,900

TYPE XXI 3817 — REF. 3817ST/X2/3ZU

Movement: automatic-winding 584Q/2 caliber; 13 ½ lines; 48-hour power reserve; 26 jewels; silicon balance spring.
Functions: hours, minutes; small seconds at 9; date at 6; chronograph: 12-hour counter at 6, 30-minute counter at 3, central seconds hand; day/night indicator at 3.
Case: steel; Ø 42mm; bidirectional rotating bezel; screw-down crown; delicately fluted caseband; sapphire crystal caseback; water resistant to 10atm.
Dial: slate gray; hours chapter with luminescent Arabic numerals; luminescent hands and dots; signed Breguet.
Strap: calfskin leather.
Suggested price: $13,900

HERITAGE 5410 — REF. 5410BR/12/9V6

Movement: automatic-winding 516GG caliber; 11 ½ lines; 65-hour power reserve; 30 jewels; silicon balance spring.
Functions: hours, minutes; small seconds at 6; date at 12.
Case: 18K rose gold; 45x32mm; delicately fluted caseband; water resistant to 3atm.
Dial: silvered 18K gold; hours chapter with luminescent Roman numerals; blued steel luminescent Breguet open-tipped hands; individually numbered and signed Breguet.
Strap: alligator leather.
Suggested price: $27,700
Also available: 18K white-gold case (ref. 5410BB/12/9V6).

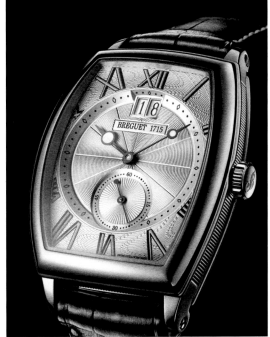

BREITLING
1884

FIRST AMONG EQUALS

Though Breitling was founded in Saint Imier in 1884, it was not until the 1940s that the watchmaker began to emblazon its name on the dial—in, fittingly enough, the Premier collection. As chronographs moved into the mainstream, **THE PIONEERING BRAND DRESSED THE COMPLICATION WITH FLAIR AND ELEGANCE, CAPTURING THE SPIRIT OF THE TIME**. New Premier models look back to those days of sure-footed style.

The flagship model of the newest Premier collection is the Premier B01 Chronograph 42. A continuation of Breitling's strong association with chronographs, this model is powered by the Breitling Manufacture Calibre 01, an automatic-winding chronometer-certified chronograph with a power reserve of more than 70 hours. Breitling was the first watchmaker to develop chronographs with separate pushers to stop and reset the timing function, a feature that persists in this model. The dial sports a clean, classic architecture that is intuitively readable, with a 30-minute counter at 3 o'clock, date window at 6 o'clock, small seconds at 9 o'clock, and a white-on-black tachometer scale around the circumference of the dial. The chronograph's 42mm case is crafted in stainless steel and features a transparent caseback that reveals the high-performance movement inside. This model is available with a dark blue or silver dial, mounted on a strap in black nubuck, alligator leather, or stainless steel.

◀ **PREMIER B01 CHRONOGRAPH 42**
The Manufacture Breitling Caliber 01 powers the stylish Premier B01 Chronograph 42 with precision and reliability.

The Premier Automatic 40 puts the clean lines and glamour of 1940s design at the service of modern machinery.

A distillation of 1940s glamour, the Premier Automatic 40 puts the clean lines of that decade at the service of modern machinery, with a nod to history in the vintage Breitling "B" at 12 o'clock. Despite its seeming simplicity, the dial includes not only a small seconds subdial at 6 o'clock, but a minute track around the periphery of the dial. Available with a blue, silver, or anthracite dial, this model is housed within a 40mm stainless steel case that is water resistant to 100m. Within beats the Breitling 37 caliber, a self-winding movement that beats at 28,800 vph and provides a minimum of 38 hours of power reserve. The retro charm of the Premier Automatic 40 benefits from the wearer's choice of a blue crocodile or brown nubuck leather strap, or seven-row stainless steel bracelet.

With an elegant curved aperture for the day of the week at 12 o'clock, and a date display at the bottom of the dial, the Premier Automatic Day and Date 40 lives up to its name with quiet confidence. Its 40mm stainless steel case follows the classic lines of the collection's original midcentury models. The sunray finish on the white dial highlights the gold coloring of the hands and indexes, radiating out toward a white minute track around the dial's circumference. The time and calendar indications are powered by the Breitling caliber 45, a certified chronometer movement that boasts 26 jewels and beats at 28,800 vph, providing a power reserve of 38 hours at a minimum. The Premier Automatic Date & Day 40 is also available with a black dial, mounted on the wearer's choice of seven-row stainless steel bracelet, or brown nubuck or crocodile leather strap.

▲ **PREMIER AUTOMATIC 40**
The stainless steel case of the Premier Automatic 40 provides the perfect masculine framework for this understated, all-business model.

▶ **PREMIER AUTOMATIC DAY & DATE 40**
Powered by the automatic-winding Breitling caliber 45, the Premier Automatic Day & Date presents the most useful calendar indications legibly and elegantly.

BREITLING

NAVITIMER 1 B01 CHRONOGRAPH 43   REF. AB0121211C1P1

Movement: automatic-winding Breitling 01 (Manufacture) caliber; 70-hour power reserve; 47 jewels; 28,800 vph.
Functions: hours, minutes; small seconds at 9; date at 4:30; chronograph: 30-minute counter at 3, 12-hour counter at 6, central 1/4-second hand.
Case: steel; Ø 43mm, thickness: 14.22mm; non screw-down crown with 2 gaskets; antireflective sapphire crystal; screwed-in sapphire caseback; water resistant to 3 bar.
Dial: blue.
Strap: black alligator leather.

NAVITIMER 1 AUTOMATIC 38   REF. U17325211G1P1

Movement: automatic-winding Breitling 17 caliber; 38-hour power reserve; 25 jewels; 28,800 vph.
Functions: hours, minutes, sweeping seconds; date at 6.
Case: steel and gold; Ø 38mm, thickness: 10.10mm; bidirectional slide rule bezel in gold; non screw-down crown with 2 gaskets; nonreflective sapphire crystal; screwed-in caseback; water resistant to 3 bar.
Dial: white.
Strap: brown alligator leather; tang buckle.

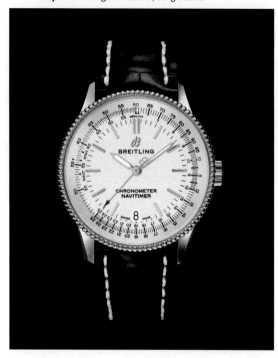

SUPEROCEAN HÉRITAGE II B20 AUTOMATIC 42   REF. AB2010161C1A1

Movement: automatic-winding Breitling 20 (Manufacture) caliber; 70-hour power reserve; 28 jewels; 28,800 vph.
Functions: hours, minutes, sweep seconds; date at 6.
Case: steel; Ø 42mm, thickness: 14.35mm; scratch-proof ceramic unidirectional ratcheted bezel; screw-down crown with two gaskets; antireflective sapphire crystal; screwed-in sapphire crystal caseback; water resistant to 200m.
Dial: blue.
Bracelet: steel mesh.

SUPEROCEAN HÉRITAGE II B01 CHRONOGRAPH 44   REF. AB0162121B1S1

Movement: automatic-winding Breitling 01 (Manufacture) caliber; 70-hour power reserve; 47 jewels; 28,800 vph.
Functions: hours, minutes; small seconds at 9; date at 4:30; chronograph: 30-minute counter at 3, 12-hour counter at 6; central 1/4-second hand.
Case: steel; Ø 44mm, thickness: 15.50mm; unidirectional ratcheted bezel; screw-down crown with two gaskets; antireflective sapphire crystal; screwed-in sapphire crystal caseback; water resistant to 200m.
Dial: black.
Strap: black rubber aero classic; push-button folding clasp.

BREITLING

NAVITIMER 8 B01 CHRONOGRAPH 43 — REF. RB0117131Q1P1

Movement: automatic-winding Breitling 01 (Manufacture) caliber; 70-hour power reserve; 47 jewels; 28,800 vph.
Functions: hours, minutes; small seconds at 9; chronograph: 30-minute counter at 3, 12-hour counter at 6, central 1/4-second hand.
Case: 18K red gold; Ø 43mm, thickness: 13.97mm; bidirectional bezel; screw-down crown with two gaskets; antireflective sapphire crystal; screwed-in sapphire crystal caseback; water resistant to 100m.
Dial: bronze.
Strap: brown alligator leather; tang buckle.

NAVITIMER 8 B01 CHRONOGRAPH 43 — REF. AB0117131B1A1

Movement: automatic-winding Breitling 01 (Manufacture) caliber; 70-hour power reserve; 47 jewels; 28,800 vph.
Functions: hours, minutes; small seconds at 9; chronograph: 30-minute counter at 3, 12-hour counter at 6, central 1/4-second hand.
Case: steel; Ø 43mm, thickness: 13.97mm; bidirectional bezel; screw-down crown with two gaskets; antireflective sapphire crystal; screwed-in sapphire crystal caseback.
Dial: black.
Bracelet: steel.

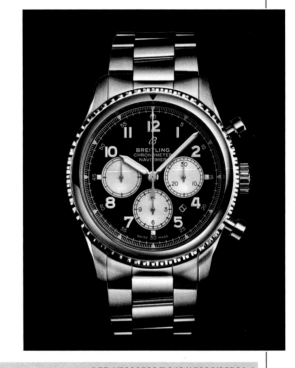

EXOSPACE B55 YACHTING — REF. EB5512221B1S1

Movement: SuperQuartz Breitling B55 (Manufacture); thermocompensated quartz electronic; analog and 12/24hr LCD digital display; backlighting (white light); battery state-of-charge indicator.
Functions: hours, minutes, seconds; countdown timer; 7 daily alarms; UTC worldtime; digital perpetual calendar with weeks indication; chronograph: 1/100th second, flyback function, electronic tachometer, lap timer, regatta timer.
Case: titanium; Ø 46mm, thickness: 15.25mm; bidirectional ratcheted bezel; non-screw-down crown with two gaskets and integrated pushpiece; antireflective sapphire crystal; screw-down caseback; water resistant to 100m.
Dial: black.
Strap: blue TwinPro rubber; pushbutton folding clasp.

EMERGENCY — REF. V7632522/BC46/156S/V20DSA.4

Movement: Breitling 76 SuperQuartz™; thermocompensated quartz electronic; analog and 12/24 hr LCD digital display; EOL indicator.
Functions: hours, minutes; countdown timer; 2nd time zone; alarm; dual-frequency distress beacon.
Case: black titanium; Ø 51mm, thickness: 21.60mm; non screw-down crown with two gaskets; bidirectional bezel with compass scale; antireflective sapphire crystal; screw-down caseback; water resistant to 5 bar.
Dial: black.
Strap: black DiverPro II rubber; pushbutton folding clasp.

BVLGARI
SLIMMER THAN EVER

The Octo Finissimo collection provides **A VERSATILE CANVAS ON WHICH TO EXPLORE WATCHMAKING CONCEPTS AND STRIP THE AESTHETICS DOWN TO BARE ESSENTIALS**. Its signature octagonal shape, inspired by architectural elements of the ancient Roman Basilica of Maxentius, frames a host of innovations, particularly relating to the quest for ever-thinner movements.

BVLGARI

With its Octo Finissimo line, Bulgari produces the thinnest tourbillons on the market.

The Finissimo models from Bulgari consistently push the limits of attainable slenderness, and the Octo Finissimo Automatic stakes out new ground. The automatic-winding Manufacture BVL 138 movement within boasts a diameter of 36.6mm with a thickness of just 2.23mm. The titanium case itself is hardly larger, measuring 40mm in diameter and 5.15mm thick. This slender volume houses a sophisticated mechanism with 36 jewels, which beats at 21,600 vph and provides a power reserve of 60 hours. With a gray dial that barely shades darker than the titanium bracelet and case, the aesthetic is one of extreme understatement. Bulgari's designers have added a grace note in the form of a small seconds subdial at 7 o'clock on the dial.

Clad in masculine gray from dial to folding buckle, the Octo Finissimo Tourbillon highlights the dynamic allure of its namesake complication with a cutout on the dial. At just 1.95mm thick, the manual-winding BVL 268 movement earns Bulgari the distinction of producing the thinnest tourbillons on the market. The sandblasted finishing of the dial, case, and bracelet—all in titanium—present a solid, cohesive visual statement. The rugged-looking 40mm case is water resistant to 30m, shielding the tourbillon with anti-reflective sapphire crystal.

BVLGARI

Within equally slim confines, the Octo Finissimo Tourbillon Automatic leaves understatement behind, revealing its inner workings with pride via a skeletonized, sandblasted titanium dial. The Finissimo Tourbilllon Automatic celebrates the beauty of the watch's functional elements, as well as the brand's own virtuosity. The wearer can admire the BVL 288 caliber, the thinnest watch in the world, in all its record-breaking glory, complete with an ultra-thin flying tourbillon at 6 o'clock on the dial. The movement itself is just 1.95mm thick, housed within a titanium case 42mm in diameter and 3.95mm thick. The titanium bracelet completes the stylish look of this exceptional piece of machinery.

▶ **OCTO FINISSIMO TOURBILLON AUTOMATIC**
The monochrome color scheme of the Octo Finissimo Tourbillon Automatic highlights its superlative technical sophistication and slimness.

THE THINNEST WATCH IN THE WORLD

BVLGARI

Bulgari turns its attention to the minute repeater, considered to be one of the most difficult complications to master because of the aural demands placed upon the mechanism. Bulgari's Octo Finissimo Minute Repeater multiplies the difficulty exponentially by shaving down its size. At 6.85mm thick, it is the thinnest minute repeater on the market, and its design is completely revolutionary. As modern in style as in construction, the model uses skeletonized indexes on the gray dial to enhance the acoustics of the sonorous complication, with the sandblasted 40mm titanium case and bracelet continuing the theme of quiet excellence. The BVL 362 caliber that powers this superb piece is visible through the transparent caseback.

▼ **OCTO FINISSIMO MINUTE REPEATER**
The limited-edition Octo Finissimo Minute Repeater boasts a 40mm case that is water resistant to 50m, an especially impressive accomplishment for a minute repeater.

BVLGARI

LVCEA — REF. 102953

Movement: automatic-winding.
Functions: hours, minutes, seconds; date.
Case: steel; Ø 33mm; antireflective crystal; water resistant to 50m.
Dial: black; diamond indexes (0.22 ct).
Bracelet: steel.

LVCEA — REF. 102954

Movement: automatic-winding.
Functions: hours, minutes, seconds; date.
Case: steel and 18K pink gold; Ø 33mm; antireflective crystal; water resistant to 50m.
Dial: mother-of-pearl; diamond indexes (0.24 ct).
Bracelet: steel and 18K pink gold.

LVCEA — REF. 103034

Movement: automatic-winding.
Functions: hours, minutes, seconds; date.
Case: 18K pink gold; Ø 33mm; set with diamonds (1.70 cts); antireflective crystal; water resistant to 50m.
Dial: mother-of-pearl; diamond indexes.
Bracelet: 18K pink gold.

LVCEA — REF. 102980

Movement: automatic-winding.
Functions: hours, minutes, seconds; date.
Case: steel and 18K pink gold; Ø 33mm; set with diamonds (1.38 cts); antireflective crystal; water resistant to 50m.
Dial: blue; diamond indexes.
Bracelet: steel and 18K pink gold.

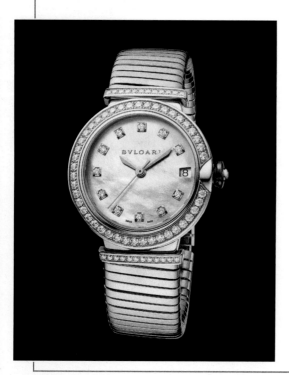

BVLGARI

LVCEA — REF. 103029

Movement: automatic-winding.
Functions: hours, minutes, seconds; date.
Case: steel and 18K pink gold; Ø 33mm; set with diamonds (1.38 cts); antireflective crystal; water resistant to 50m.
Dial: anthracite; diamond indexes.
Bracelet: steel and 18K pink gold.

LVCEA — REF. 102833

Movement: automatic-winding.
Functions: hours, minutes; seconds.
Case: 18K pink gold; Ø 30mm; set with diamonds (1.72 cts); water resistant to 50m.
Dial: skeletonized; BVLGARI spelled out in 18K pink gold and inlaid with diamonds.
Strap: red alligator leather; tang buckle.

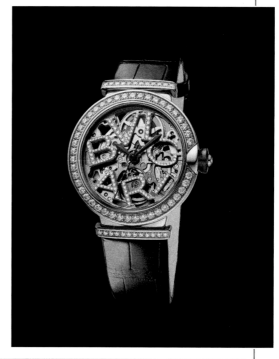

DIVAS' DREAM — REF. 102842

Movement: automatic-winding.
Functions: hours, minutes.
Case: 18K white gold; Ø 37mm; set with diamonds (2.38 cts) and sapphires (0.98 ct); water resistant to 30m.
Dial: blue speckled with 18K white gold; set with diamonds and sapphires.
Strap: blue alligator leather; deployant clasp.

DIVAS' DREAM — REF. 102843

Movement: automatic-winding.
Functions: hours, minutes.
Case: 18K pink gold; Ø 37mm; set with diamonds (2.38 cts) and sapphires (0.98 ct); water resistant to 30m.
Dial: blue speckled with 18K pink gold; set with diamonds and sapphires.
Strap: blue alligator leather; deployant clasp.

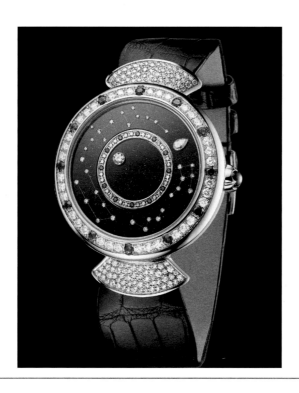

BVLGARI

LVCEA — REF. 103039

Movement: automatic-winding.
Functions: hours, minutes; tourbillon.
Case: 18K white gold; Ø 38mm; set with diamonds (5.01 cts); antireflective crystal; water resistant to 50m.
Dial: snow-set diamonds (1.68 cts).
Strap: galuchat.
Note: limited edition of 25 pieces.

LVCEA — REF. 102881

Movement: automatic-winding.
Functions: hours, minutes; tourbillon.
Case: 18K white gold; Ø 38mm; set with diamonds (5.01 cts); antireflective crystal; water resistant to 50m.
Dial: blue; with snow-set diamonds (1.68 cts).
Strap: galuchat; deployant buckle.
Note: limited edition of 25 pieces.

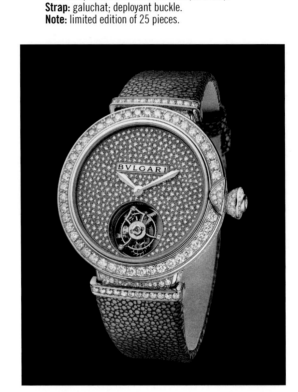

LVCEA — REF. 102887

Movement: automatic-winding.
Functions: hours, minutes; tourbillon.
Case: 18K white gold; Ø 38mm; set with diamonds (5.01 cts); antireflective crystal; water resistant to 50m.
Dial: pink; with snow-set diamonds (1.68 cts).
Strap: galuchat; deployant buckle.
Note: limited edition of 25 pieces.

DIVA FINISSIMA REPETITION MINUTES — REF. 102839

Movement: manual-winding BVL 362 caliber; thickness: 1.95mm; 52-hour power reserve.
Functions: hours, minutes; small seconds at 6; minute repeater.
Case: 18K pink gold; Ø 37mm; set with brilliant-cut diamonds (~4.26 cts); 18K pink-gold crown set with a faceted diamond; sapphire crystal caseback; water resistant to 3atm.
Dial: black lacquer speckled with gold dust using the Japanese artisanal Urushi technique; hour markers and small seconds set with brilliant-cut diamonds (~0.28 ct).
Strap: black alligator leather; 18K pink-gold folding clasp set with diamonds (~0.29 ct).
Note: limited edition of 10 pieces.

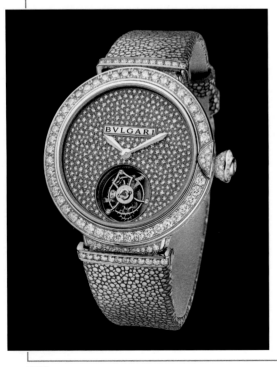

BVLGARI

OCTO FINISSIMO — REF. 102711

Movement: automatic-winding Manufacture BVL 138 Finissimo Caliber; thickness: 2.23mm; 60-hour power reserve; 36 jewels; 21,600 vph.
Functions: hours, minutes; small seconds.
Case: titanium; Ø 40mm.
Dial: gray.
Strap: black alligator leather; tang buckle.

OCTO FINISSIMO — REF. 102912

Movement: automatic-winding Manufacture BVL 148 Finissimo caliber; thickness: 2.23mm; 60-hour power reserve; 36 jewels; 21,600 vph.
Functions: hours, minutes; small seconds.
Case: 18K pink gold; Ø 40mm; antireflective crystal; water resistant to 30m.
Dial: pink-gold colored.
Bracelet: 18K pink-gold bracelet; deployant strap.

OCTO FINISSIMO — REF. 102714

Movement: manual-winding BVL 128SK caliber; Ø 36mm, thickness: 2.35mm; 65-hour power reserve; 28 jewels; 28,800 vph; openworked and blackened mainplate and bridges decorated with circular brushing and chamfered finishing.
Functions: hours, minutes; small seconds between 7 and 8; power reserve indicator at 9.
Case: sand-blasted titanium; Ø 40mm, thickness: 5.37mm; screw-down titanium crown with ceramic insert; scratch-resistant antireflective sapphire crystal; sapphire crystal caseback; water resistant to 3atm.
Dial: skeletonized; minute, second and power reserve tracks; faceted and skeletonized rhodium-plated hands.
Strap: black alligator leather; titanium ardillon buckle.

OCTO FINISSIMO — REF. 102946

Movement: manual-winding BVL 128SK caliber; Ø 36mm, thickness: 2.35mm; 65-hour power reserve; 28 jewels; 28,800 vph; openworked and blackened mainplate and bridges decorated with circular brushing and chamfered finishing.
Functions: hours, minutes; small seconds between 7 and 8; power reserve indicator at 9.
Case: 18K pink gold; Ø 40mm, thickness: 5.37mm; screw-down 18K pink-gold crown with ceramic insert; scratch-resistant antireflective sapphire crystal; sapphire crystal caseback; water resistant to 3atm.
Dial: skeletonized; minute, second and power reserve tracks; faceted and skeletonized 18K pink-gold-plated hands.
Strap: black alligator leather; 18K pink-gold ardillon buckle.

BVLGARI

OCTO FINISSIMO — REF. 103010

Movement: manually-wound BVL 128SK Finissimo Caliber; thickness: 2.35mm; 65-hour power reserve; 28 jewels; 28,800 vph.
Functions: hours, minutes, small seconds; power reserve indicator.
Case: titanium treated DLC; Ø 40mm; antireflective crystal; water resistant to 30m.
Dial: black.
Bracelet: titanium treated DLC.

OCTO FINISSIMO — REF. 102794

Movement: manually-wound.
Functions: hours, minutes; seconds; minute repeater.
Case: carbon-titanium; Ø 40mm; water resistant to 10m.
Dial: anthracite.
Strap: black carbon.
Bracelet: black carbon.

OCTO FINISSIMO — REF. 102560

Movement: manually-wound Manufacture BVL 268 Caliber FinissimoTourbillon; thickness: 1.95mm; 52-hour power reserve; 11 jewels; 21,600 vph.
Functions: hours, minutes; seconds at 6; flying tourbillon.
Case: black titanium treated DLC; Ø 32.60mm; water resistant to 30m.
Dial: black; applied stick markers.
Strap: black alligator leather; tang buckle.
Note: the thinnest tourbillon mechanical watch in the world.

OCTO FINISSIMO TOURBILLON — REF. 102138

Movement: manual-winding BVL 268 caliber; Ø 32.6mm, thickness: 1.95mm; 52-hour power reserve; 13 jewels; 21,600 vph.
Functions: hours, minutes; tourbillon at 6.
Case: platinum; Ø 40mm, thickness: 5mm; crown with black ceramic insert; scratch-resistant antireflective sapphire crystal; sapphire crystal caseback.
Dial: black lacquered; faceted and skeletonized hands.
Strap: black alligator leather; platinum ardillon buckle.
Note: the thinnest tourbillon mechanical watch in the world.

BVLGARI

SERPENTI — REF. 102886

Movement: quartz.
Functions: hour, minutes.
Case: white ceramic; Ø 35mm; 18K pink-gold bezel set with diamonds (0.29 ct); water resistant to 30m.
Dial: white; metal indexes.
Bracelet: white ceramic; double spiral.

SERPENTI — REF. 102885

Movement: quartz.
Functions: hour, minutes.
Case: black ceramic; Ø 35mm; 18K pink-gold bezel set with diamonds (0.29 ct); water resistant to 30m.
Dial: black; metal indexes.
Bracelet: black ceramic; double spiral.

SERPENTI — REF. 102948

Movement: quartz.
Functions: hours, minutes.
Case: 18K pink gold; Ø 35mm; 18K pink-gold bezel set with diamonds (0.29ct); 18K pink-gold crown set with a cabochon-cut rubellite; water resistant to 30m.
Dial: black; metal indexes.
Bracelet: 18K yellow, pink, and white golds; double spiral.

SERPENTI — REF. 103002

Movement: quartz.
Functions: hours, minutes.
Case: 18K pink gold; Ø 35mm; 18K pink-gold bezel set with diamonds (0.29 ct); 18K pink-gold crown set with a cabochon-cut rubellite; water resistant to 30m.
Dial: white; metal indexes.
Bracelet: 18K pink gold; double spiral.

BVLGARI

SERPENTI   REF. 102728

Movement: quartz.
Functions: hours, minutes.
Case: curved 18K pink gold; Ø 27mm; 18K pink-gold crown set with a cabochon-cut rubellite.
Dial: black; pink-gold-plated hands, indexes.
Bracelet: 18K pink gold.

SERPENTI MISTERIOSI   REF. 102784

Movement: quartz.
Functions: hours, minutes.
Case: 18K white gold; Ø 40mm; antireflective crystal; water resistant to 30m.
Dial: pavé-set.
Bracelet: 18K white gold; set with brilliant-cut diamonds.
Note: bracelet and case set with brilliant-cut diamonds (total: 21.58 cts) and emeralds (total: 4.18 cts).

SERPENTI   REF. 102446

Movement: quartz.
Functions: hours, minutes.
Case: 18K pink gold; Ø 40mm; set with pavé diamonds (4.95 cts) and two malachites; antireflective crystal; water resistant to 30m.
Dial: blue lacquer and diamonds.
Bracelet: 18K pink gold; double spiral: coated with blue and green lacquer.

SERPENTI   REF. 102701

Movement: quartz.
Functions: hours, minutes.
Case: 18K white gold; Ø 40mm; set with brilliant-cut diamonds and 2 pear-shaped emeralds (0.4 ct); scratch-resistant antireflective sapphire crystal; water resistant to 3atm.
Dial: 18K white gold; set with brilliant-cut diamonds; green-coated brass hands.
Bracelet: 18K white gold; double spiral; set with brilliant-cut diamonds.
Note: set with 963 diamonds total (33.37 carats).

BVLGARI

SERPENTI — REF. 102981

Movement: quartz.
Functions: hours, minutes.
Case: 18K pink gold; Ø 36mm; antireflective crystal; water resistant to 30m.
Dial: black lacquer.
Bracelet: 18K pink gold; black lacquer; set with 2 pear-shaped amethysts (0.47 ct).

SERPENTI — REF. 102982

Movement: quartz.
Functions: hours, minutes.
Case: 18K pink gold; Ø 36mm; antireflective crystal; water resistant to 30m.
Dial: mother-of-pearl.
Bracelet: 18K pink gold; set with pavé diamonds (0.73 ct) and 2 pear-shaped amethysts (0.5 ct).

SERPENTI — REF. 102983

Movement: quartz.
Functions: hours, minutes.
Case: 18K pink gold; Ø 36mm; antireflective crystal; water resistant to 30m.
Dial: pavé-set diamonds.
Bracelet: 18K pink gold; set with pavé diamonds (2.79 cts) and 2 pear-shaped amethysts (0.51 ct).

SERPENTI — REF. 102989

Movement: quartz.
Functions: hours, minutes.
Case: 18K white gold; Ø 36mm; antireflective crystal; water resistant to 30m.
Dial: pavé-set diamonds.
Bracelet: 18K white gold; set with pavé diamonds (11.53 cts) and 2 pear-shaped sapphires (0.65 ct).

CARL F. BUCHERER
LUCERNE 1888

AROUND THE EDGES

Many haute horology timepieces feature a transparent caseback, the better to showcase the mastery and dedication that goes into Swiss made calibers. With a traditional rotor, however, a portion of the movement is covered at every moment. Carl F. Bucherer has pioneered **THE PERIPHERAL WINDING ROTOR, WHICH IS PLACED AT THE EDGE OF THE MOVEMENT, REVEALING THE WORKMANSHIP THROUGHOUT**. Two new models featuring in-house movements explore the possibilities of this unconventional technique.

The timeless style of Carl F. Bucherer's Manero Tourbillon Double Peripheral expresses an understated confidence in the expertise within. Moving the tourbillon from its usual position at 6 o'clock to the top of the dial corresponds with the emphasis on technical brilliance in the mechanism's construction. The 43mm rose-gold case provides a smooth and sleek home for the new groundbreaking CFB T3000 movement that powers the watch, and its characteristic shape reveals its solid standing as a proud member of the Manero family. The convex silver-colored dial continues the visual theme of gently rounded forms made for prime wearability. Discreet rose-gold-plated indexes and hands match the small seconds display on the tourbillon itself, which features a stop seconds function. Mounted on a hand-stitched brown alligator leather strap, this model exudes dignified luxury.

◀ **MANERO TOURBILLON DOUBLE PERIPHERAL**

The smooth, elegant design of the Manero Tourbillon Double Peripheral highlights the exceptional peripherally-mounted tourbillon at the top of the dial.

CARL F. BUCHERER

The extraordinary CFB T3000 creates a peripheral support for the tourbillon carriage, thus revealing it completely from the dial and the caseback.

Behind the understated exterior, the model's powerhouse within provides a tour de force. Carl F. Bucherer specializes in peripheral winding movements, and the CFB T3000 applies these principles to the construction of the tourbillon—long the epitome of watchmaking sophistication. Most tourbillons are mounted on the watch's mainplate and the balance-wheel bridge, with the specialized "flying tourbillon" mounted only to the mainplate. The extraordinary CFB T3000 goes above and beyond, creating a peripheral support for the tourbillon carriage using three ball bearings. The tourbillon is thus completely visible from the dial and the caseback, seeming to float weightlessly within the timepiece. The movement is just as modern in its construction as it is in its design, with the escapement's pallet and escape wheel crafted in silicon. Such a material is impervious to magnetic fields and requires no lubrication, leading to a power reserve of at least 65 hours. Carl F. Bucherer's characteristic peripheral winding rotor leaves the view of the movement through the sapphire caseback completely unhindered. This technique hides the supportive ball bearings, but shows off the Côtes de Genève finishing on the movement's bridge, a nod to Swiss watchmaking tradition. The CFB T3000 includes two patented devices from Carl F. Bucherer: one for the rotor, and one for the tourbillon. Both use ball bearings for peripheral support. The movement is one of the few tourbillon movements to boast the distinction of being a COSC-certified chronometer.

▶ **CFB T3000 MOVEMENT**

Comprising 189 parts, the CFB T3000 movement uses groundbreaking technology to provide a generous power reserve, first-class precision, and a stunning view of the tourbillon and the rest of the movement.

CARL F. BUCHERER

◀ **MANERO PERIPHERAL**
The rose-gold case of this Manero Peripheral is the perfect complement to its light brown alligator leather strap.

Enlarging the case of the Manero Peripheral, Carl F. Bucherer provides the CFB A2050 movement with a more capacious frame. The new 43mm case maintains its sense of perfect proportion, with simple appliqué wedge-shaped indexes continuing on a path of visual equilibrium. The warm color scheme of the new Manero model combines rose gold and a light brown alligator leather strap, creating a counterpoint to the simple white dial. Water resistant to 50m, the models bears sapphire crystal on case front and back. A date aperture at 3 o'clock and a small seconds display at 6 o'clock round out the understated face.

▲ **A2000 MOVEMENT**
Designed to be the foundation for an array of calibers, the base A2000 movement relies on its peripheral winding rotor, which reveals the rest of the movement at any given time.

In developing the CFB A2000 movement, Carl F. Bucherer has made a huge stride towards realizing many of its future horological achievements. The foundation of an entire series of movements, the CFB A2000 has the flexibility and range to add functions and features as desired by the watchmakers. It derives its power from a double-sided automatic-winding mechanism equipped with a peripheral rotor. The caliber CFB A1000 was the first movement to exploit this innovative arrangement, and CFB A2000 carries forward the baton, while keeping in place elements such as the Côtes de Genève of the initial caliber's bridges and balance-cocks. Carl F. Bucherer's watchmaking excellence also entails an emphasis on high accuracy with the use of pivotable masselottes, small weights that control the inertia of the balance. The Manero Peripheral is the first collection to welcome the CFB A2000 into its ranks, in the CFB A2050, an attractively laid out variation that powers hours, minutes, small seconds, and the date at 3 o'clock. It powers the Manero Peripheral at 28,800 vph with precision and finesse.

CARL F. BUCHERER

MANERO FLYBACK — REF. 00.10919.08.33.01

Movement: automatic-winding CFB 1970 caliber; Ø 30.4mm, thickness: 7.9mm; 42-hour power reserve; 25 jewels.
Functions: hours, minutes; small seconds at 9; date at 6; flyback chronograph: central seconds hand, 30-minute counter at 3; tachometer scale.
Case: stainless steel; Ø 43mm, thickness: 14.45mm; antireflective double-domed sapphire crystal; sapphire crystal caseback; water resistant to 3atm.
Dial: black.
Strap: Lousiana alligator leather; stainless steel folding pin buckle.

Suggested price: $6,200
Also available: silver dial (ref. 00.10919.08.13.01).

MANERO FLYBACK — REF. 00.10919.08.33.02

Movement: automatic-winding CFB 1970 caliber; Ø 30.4mm, thickness: 7.9mm; 42-hour power reserve; 25 jewels.
Functions: hours, minutes; small seconds at 9; date at 6; flyback chronograph: central seconds hand, 30-minute counter at 3; tachometer scale.
Case: stainless steel; Ø 43mm, thickness: 14.45mm; antireflective double-domed sapphire crystal; sapphire crystal caseback; water resistant to 3atm.
Dial: black; silver subdials.
Strap: Kudu leather; stainless steel folding pin buckle.

Suggested price: $6,200

MANERO FLYBACK — REF. 00.10919.03.33.01

Movement: automatic-winding CFB 1970 caliber; Ø 30.4mm, thickness: 7.9mm; 42-hour power reserve; 25 jewels.
Functions: hours, minutes; small seconds at 9; date at 6; flyback chronograph: central seconds hand, 30-minute counter at 3; tachometer scale.
Case: 18K rose gold; Ø 43mm, thickness: 14.45mm; antireflective double-domed sapphire crystal; sapphire crystal caseback; water resistant to 3atm.
Dial: black.
Strap: Louisiana alligator leather; 18K rose-gold pin buckle.

Suggested price: $16,900
Also available: silver dial (ref. 00.10919.03.13.01); champagne dial (ref. 00.10919.03.43.01).

MANERO FLYBACK — REF. 00.10919.08.93.01

Movement: automatic-winding CFB 1970 caliber; Ø 30.4mm, thickness: 7.9mm; 42-hour power reserve; 25 jewels.
Functions: hours, minutes; small seconds at 9; date at 6; flyback chronograph: central seconds hand, 30-minute counter at 3; tachometer scale.
Case: stainless steel; Ø 43mm, thickness: 14.45mm; antireflective double-domed sapphire crystal; sapphire crystal caseback; water resistant to 3atm.
Dial: blue-gray.
Strap: Louisiana alligator leather; stainless steel folding pin buckle.

Suggested price: $6,200

CARL F. BUCHERER

MANERO PERIPHERAL — REF. 00.10921.03.23.01

Movement: automatic-winding CFB A2050 Manufacture caliber; Ø 30.6mm, thickness: 5.28mm; 55-hour power reserve; 33 jewels; chronometer certified.
Functions: hours, minutes; small seconds at 6; date at 3.
Case: 18K rose gold; Ø 43.1mm, thickness: 11.2mm; antireflective convex sapphire crystal; sapphire crystal caseback; water resistant to 3atm.
Dial: white.
Strap: Louisiana alligator leather; 18K rose-gold pin buckle.
Suggested price: $17,600
Also available: black dial (ref. 00.10921.03.33.01).

MANERO PERIPHERAL — REF. 00.10921.08.23.01

Movement: automatic-winding CFB A2050 Manufacture caliber; Ø 30.6mm, thickness: 5.28mm; 55-hour power reserve; 33 jewels; chronometer certified.
Functions: hours, minutes; small seconds at 6; date at 3.
Case: stainless steel; Ø 43.1mm, thickness: 11.2mm; antireflective convex sapphire crystal; sapphire crystal caseback; water resistant to 3atm.
Dial: white.
Strap: Louisiana alligator leather; stainless steel folding pin buckle.
Suggested price: $6,800
Also available: stainless steel bracelet (ref. 00.10921.08.23.21).

MANERO PERIPHERAL — REF. 00.10921.08.33.01

Movement: automatic-winding CFB A2050 Manufacture caliber; Ø 30.6mm, thickness: 5.28mm; 55-hour power reserve; 33 jewels; chronometer certified.
Functions: hours, minutes; small seconds at 6; date at 3.
Case: stainless steel; Ø 43.1mm, thickness: 11.2mm; antireflective convex sapphire crystal; sapphire crystal caseback; water resistant to 3atm.
Dial: black.
Strap: Louisiana alligator leather; stainless steel folding pin buckle.
Suggested price: $6,800
Also available: stainless steel bracelet (ref. 00.10921.08.33.21).

MANERO PERIPHERAL — REF. 00.10917.08.53.99

Movement: automatic-winding CFB A2050 Manufacture caliber; Ø 30.6mm, thickness: 5.28mm; 55-hour power reserve; 33 jewels; chronometer certified.
Functions: hours, minutes; small seconds at 6; date at 3.
Case: stainless steel; Ø 40.6mm, thickness: 11.2mm; antireflective domed sapphire crystal; sapphire crystal caseback; water resistant to 3atm.
Dial: blue.
Strap: Louisiana alligator leather; stainless steel folding pin buckle.
Suggested price: $6,800

CARL F. BUCHERER

PATRAVI TRAVELTEC   REF. 00.10620.08.33.02

Movement: automatic-winding CFB 1901.1 caliber; Ø 28.6mm, thickness: 7.3mm; 42-hour power reserve; 39 jewels; chronometer certified.
Functions: hours, minutes; small seconds at 3; date between 4 and 5; three time zones; chronograph: central seconds hand, 12-hour counter at 6, 30-minute counter at 3.
Case: stainless steel; Ø 46.6mm, thickness: 15.5mm; screw-down crown; antireflective sapphire crystal; water resistant to 5atm.
Dial: black.
Strap: rubber; stainless steel folding clasp with comfort extension.
Suggested price: $10,900
Also available: stainless steel bracelet (ref. 00.10620.08.33.21).

PATRAVI TRAVELTEC   REF. 00.10620.03.33.01

Movement: automatic-winding CFB 1901.1 caliber; Ø 28.6mm, thickness: 7.3mm; 42-hour power reserve; 39 jewels; chronometer certified.
Functions: hours, minutes; small seconds at 3; date between 4 and 5; three time zones; chronograph: central seconds hand, 12-hour counter at 6, 30-minute counter at 3.
Case: 18K rose gold; Ø 46.6mm, thickness: 15.5mm; screw-down crown; antireflective sapphire crystal; water resistant to 5atm.
Dial: black.
Strap: calfskin leather; 18K rose-gold pin lock folding clasp.
Suggested price: $41,800
Also available: 18K rose-gold bracelet (ref. 00.10620.03.33.21).

PATRAVI SCUBATEC   REF. 00.10632.22.33.01

Movement: automatic-winding CFB 1950.1 caliber; Ø 26.2mm, thickness: 4.6mm; 38-hour power reserve; 25 jewels; chronometer certified.
Functions: hours, minutes, seconds; date at 3; automatic helium release valve.
Case: 18K rose gold and blackened titanium; Ø 44.6mm, thickness: 13.45mm; 18K rose-gold and ceramic bezel; screw-down crown; antireflective sapphire crystal; water resistant to 50atm.
Dial: black.
Strap: rubber; 18K rose gold and blackened titanium folding clasp with comfort extension.
Suggested price: $23,600
Also available: blue dial (ref. 00.10632.22.53.01).

PATRAVI SCUBATEC   REF. 00.10632.24.53.01

Movement: automatic-winding CFB 1950.1 caliber; Ø 26.2mm, thickness: 4.6mm; 38-hour power reserve; 25 jewels; chronometer certified.
Functions: hours, minutes, seconds; date at 3; automatic helium release valve.
Case: stainless steel; Ø 44.6mm, thickness: 13.45mm; 18K rose-gold and ceramic bezel; screw-down crown; antireflective sapphire crystal; water resistant to 50atm.
Dial: blue.
Strap: rubber; stainless steel folding clasp with comfort extension.
Suggested price: $9,600
Also available: black dial (ref. 00.10632.24.33.01); white dial (ref. 00.10632.24.23.01).

CARL F. BUCHERER

ALCARIA SWAN   REF. 00.10702.02.90.27

Movement: quartz CFB 1850 caliber.
Functions: hours, minutes.
Case: 18K white gold; 26.5x38mm, thickness: 7.4mm; set with 348 FC TW vvs diamonds (3.3 carats); antireflective domed sapphire crystal; water resistant to 3atm.
Dial: 18K white gold; set with 137 FC TW vvs diamonds (1.2 carats).
Bracelet: 18K white gold; set with 844 FC TW vvs diamonds (8.7 carats).
Note: limited edition of 88 pieces.
Suggested price: $158,000

ALCARIA QUEEN   REF. 00.10701.07.15.31

Movement: quartz CFB 1850 caliber.
Functions: hours, minutes.
Case: stainless steel and 18K rose gold; 26.5x38mm, thickness: 7.4mm; set with 38 FC TW vvs diamonds (0.6 carat); antireflective domed sapphire crystal; water resistant to 3atm.
Dial: silver; 18K rose-gold Roman numerals.
Bracelet: stainless steel and 18K rose gold.
Suggested price: $11,200

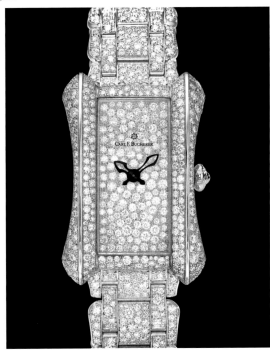

PATHOS SWAN   REF. 00.10590.03.90.31

Movement: quartz CFB 1851 caliber.
Functions: hours, minutes.
Case: 18K rose gold; Ø 34mm, thickness: 9.45mm; set with 254 FC TW vvs diamonds and 116 sapphires (4.15 carats); antireflective sapphire crystal; water resistant to 3atm.
Dial: 18K rose gold and mother-of-pearl; set with 31 FC TW vvs diamonds and 117 sapphires (0.6 carat).
Bracelet: 18K rose gold; set with 240 FC TW vvs diamonds and 164 sapphires (7.75 carats); folding clasp.
Note: limited edition of 88 pieces.
Suggested price: $162,000

PATHOS DIVA   REF. 00.10580.08.23.31.02

Movement: automatic-winding CFB 1963 caliber; Ø 20mm, thickness: 4.8mm; 38-hour power reserve; 25 jewels.
Functions: hours, minutes, seconds; date at 3.
Case: stainless steel; Ø 34mm, thickness: 9.65mm; set with 54 FC TW vvs diamonds (0.7 carat); antireflective sapphire crystal; water resistant to 3atm.
Dial: white.
Bracelet: stainless steel; folding clasp.
Suggested price: $8,600

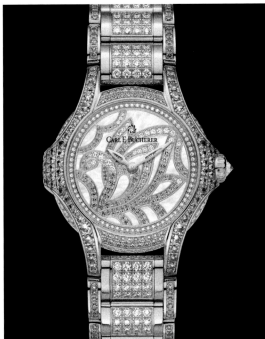

Cartier

THE MORE THINGS CHANGE...

With a history over a century long, **CARTIER HAS DEVELOPED AN AESTHETIC VOCABULARY THAT IS UNUSUALLY DEEP AND RICH.** Where most brands would be thrilled to count one immediately recognizable signature motif among their collections, Cartier boasts a plethora. Some are endlessly mutable, changing with the times while staying true to an essential spirit, while some still sport the visual trademarks of the collection's very first model, as fresh today as in the beginning.

◀ **BAIGNOIRE ALLONGÉE MOYEN MODÈLE**
The unusually shaped rose-gold case of the Baignoire Allongée Moyen Modèle measures 47x21mm, with 8mm of thickness, and provides water resistance to approximately 30m, as well as a dramatic beaded look.

▶ **BAIGNOIRE PETIT MODÈLE**
Pavé-set with brilliant-cut diamonds, the 18-karat white-gold case of the Baignoire Petit Modèle houses a quartz movement.

The eye-catching curves of the Baignoire collection first joined Cartier's ranks in 1912, when Louis Cartier began playing with the traditional round watch shape.

The shapes of cases have always been an intriguing variable in the world of high-end watches, not least because they often present a technical challenge in the form of an aesthetic one, given the new considerations and constraints placed on the movement, the hands, the strap, and even the wearer. First introduced during the rule-breaking, devil-may-care "Swinging London" of the 1960s, the Baignoire Allongée stretched the oval case to extreme proportions. This Baignoire Allongée Moyen Modèle boasts dimensions of 47x21mm, making a dramatic adornment for any wrist—even before the addition of dramatic gold beads, reminiscent of rounded pyramids or reptilian armor, on the 18-karat rose-gold case. The beads also adorn the crown, which is set with a brilliant-cut diamond. Blued steel sword-shaped hands traverse the silvered dial, whose Roman numerals stretch and compress to accommodate the unusual case dimensions. Within beats the manual-winding 1917 MC caliber, created by Cartier, which provides the wearer with a power reserve of 38 hours. Completing this picture of bold femininity, the gray alligator leather strap and rose-gold ardillon buckle lend just the right note of sobriety to the gleaming case and dial.

The eye-catching curves of the Baignoire collection first joined Cartier's ranks in 1912, when Louis Cartier began playing with the traditional round watch shape, endlessly experimenting with form and function. The size and shape of the Baignoire Petit Modèle could make for a more understated experience, but the sparkle of dozens of thickly set brilliant-cut diamonds on the 18-karat white-gold case cannot fail to catch the eye. The case exhibits the collection's characteristic oval shape ("baignoire" meaning "bathtub" in French), with dimensions of 32x26mm, and 8mm in height. A brilliant-cut diamond adorns the beaded crown for an additional point of light. The blued steel sword-shaped hands and silvered dial conform perfectly to the grown-up elegance of the model's color scheme, constructed as it is in gleaming white and muted blue, with the grace note of the blue alligator leather strap and white-gold ardillon buckle.

CARTIER

Famed aviator Alberto Santos-Dumont was the first person to wear a wristwatch while flying—custom-made for him by his good friend Louis Cartier. A pioneer in the realm of wristwatches, the Santos-Dumont collection continues a proud Cartier tradition. Its distinctive square case, crafted in stainless steel, sets an adventurous tone confirmed by its dimensions: 43.5x31.4mm, with a thickness of 7.3mm for the Grand Modèle. This shape was a revolutionary break from tradition in the early 20th century, as the world of watchmaking remained beholden to circular pocket watches. The Roman numerals, visible screws, beaded crown and blue cabochon attest to the heritage of this iconic piece. Efficiency, practicality, and ease of use were the watchwords of the early aviators, particularly the dashing Santos-Dumont. Continuing this legacy, Cartier reworked the high-efficiency quartz movement within, reducing energy consumption and adding a new high-performance battery to achieve a power reserve of approximately six years. Mounted on an alligator leather strap, the Santos-Dumont plays up its classic, nonchalant elegance while remaining effortlessly modern—much like the Eiffel Tower, whose square base is said to be an aesthetic influence.

◀ **SANTOS-DUMONT GRAND MODÈLE**
Originally designed in the early days of aviation, the Santos-Dumont Grand Modèle maintains the daring glamour of its first incarnation, with a cutting-edge interior.

CARTIER

An emblematic animal of the house of Cartier, the panther is an endless source of ideas for the brand's designers, who revel in its streak of dark, powerful femininity. Two new takes on the idea prove that the possibilities are endless.

The "manchette" model interprets the theme abstractly. The 18-karat rose-gold bracelet bears cleverly arranged "spots" of black lacquer, positioned here and there like the spots of a panther. The panther's spots are camouflage, and so too are these—it takes a moment to distinguish the 18-karat rose-gold dial, as it blends in with the wild motif of the bracelet. The golden finish on the steel sword-shaped hands adds to the illusion. Though the dimensions of the case proper are 22x19mm, the entire bracelet is 41mm wide, combining a bold, dramatic look with the diminutive volume of a traditionally feminine watch. The model throws another curveball by moving the case and dial to the extreme right of the case, its unexpected placement camouflaging it further. The quartz movement inside allows for a slimness of 6.5mm. This stunning watch is available in a limited edition of 50 individually numbered pieces.

A more representational panther prowls on the black lacquer dial of the Panthère Dentelle, glaring out with one emerald eye. Brilliant-cut diamonds provide a gleaming backdrop for the panther's black lacquer spots. Though the diameter of the case is 36mm, the time display takes up just a small portion of the dial, protected and encircled by a ring of diamonds. The panther, dwarfing the petite steel sword-shaped hands with rhodium finish, is clearly in control here, and the diamond-pavé bezel seems to just barely contain the animal's power within the 18-karat white-gold case. Versatile in its sleek elegance, the watch comes with two alligator leather straps, one in black and one in dark green. The 18-karat white-gold ardillon buckle is set with brilliant-cut diamonds, continuing the theme established on the case and dial.

▲ PANTHÈRE DE CARTIER MANCHETTE EXTRA LARGE
Borrowing its glamour and danger from Cartier's iconic feline, the Panthère de Cartier Manchette does double duty as bold jewelry and understated timepiece.

▶ PANTHÈRE DENTELLE
Within a case that measures 36mm in diameter and 8.8mm in thickness, the Panthère Dentelle gives over most of its dial space to its namesake hunter.

CARTIER

BAIGNOIRE ALLONGÉE MEDIUM MODEL — REF. HPI01306

Movement: manual-winding caliber 1917 MC; 38-hour power reserve.
Functions: hours, minutes.
Case: 18K white gold; Ø 47x21mm, thickness: 8mm; set with brilliant-cut diamonds; beaded crown set with a brilliant-cut diamond; sapphire crystal; water resistant to 3 bar.
Dial: 18K white-gold dial; set with brilliant-cut diamonds; blued steel sword-shaped hands.
Bracelet: 18K white gold; set with brilliant-cut diamonds.
Price: available upon request.

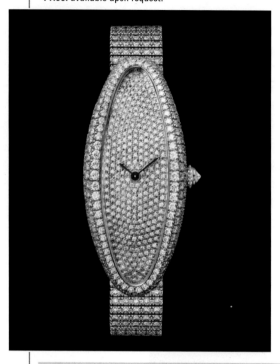

BAIGNOIRE ALLONGÉE MEDIUM MODEL — REF. WJBA0006

Movement: manual-winding caliber 1917 MC; 38-hour power reserve.
Functions: hours, minutes.
Case: 18K pink gold; Ø 47x21mm, thickness: 8mm; set with brilliant-cut diamonds; beaded crown set with a brilliant-cut diamond; sapphire crystal; water resistant to 3 bar.
Dial: silvered; blued steel sword-shaped hands.
Strap: taupe alligator leather; 18K pink-gold ardillon buckle.
Price: available upon request.

BAIGNOIRE ALLONGÉE MEDIUM MODEL — REF. WGBA0009

Movement: manual-winding caliber 1917 MC manufacture; 38-hour power reserve.
Functions: hours, minutes.
Case: 18K pink gold; Ø 47x21mm, thickness: 8mm; beaded crown set with a brilliant-cut diamond; sapphire crystal; water resistant to 3 bar.
Dial: silvered; blued steel sword-shaped hands.
Strap: gray alligator leather; 18K pink-gold ardillon buckle.
Price: available upon request.

BAIGNOIRE SMALL MODEL — REF. WGBA0007

Movement: quartz.
Functions: hours, minutes.
Case: 18K yellow gold; Ø 32x26mm, thickness: 8mm; beaded crown set with a sapphire cabochon; sapphire crystal; water resistant to 3 bar.
Dial: silvered; blued steel sword-shaped hands.
Strap: taupe alligator leather; 18K yellow-gold ardillon buckle.
Price: available upon request.

CARTIER

SANTOS DE CARTIER SQUELETTE LARGE MODEL — REF. WHSA0009

Movement: manual-winding caliber 9612 MC; 72-hour power reserve.
Functions: hours, minutes.
Case: steel ADLC; 39.8x47.5mm, thickness: 9.1mm; heptagonal crown set with a blue spinel; sapphire crystal; water resistant to 10 bar.
Dial: skeletonized bridges in the shape of Roman numerals, covered with SuperLumiNova; gray steel sword-shaped hands covered with SuperLumiNova.
Strap: black alligator leather; comes with a second gray alligator leather strap; steel ADLC deployant buckle.
Price: available upon request.

SANTOS DE CARTIER CHRONO EXTRA LARGE — REF. WSSA0017

Movement: automatic-winding caliber 1904-CH M, 48-hour power reserve.
Functions: hours, minutes; chronograph; date.
Case: steel, ADLC case; 43.3x51.3mm, thickness: 12.5mm; heptagonal crown set with a faceted synthetic spinel; sapphire crystal; water resistant to 10 bar.
Dial: silvered; satin finish; steel sword-shaped hands covered with SuperLumiNova.
Strap: black rubber and alligator leather; interchangeable straps; interchangeable steel buckle.
Price: available upon request.

SANTOS-DUMONT SMALL MODEL — REF. WGSA0022

Movement: quartz; approx. 6-year power reserve.
Functions: hours, minutes.
Case: 18K pink gold; 38x27.5mm, thickness: 7.3mm; beaded crown set with a sapphire cabochon; sapphire crystal; water resistant to 3 bar.
Dial: silvered; sunray finish; blued steel sword-shaped hands.
Strap: dark gray semi-matte alligator leather; 18K pink-gold ardillon buckle.
Price: available upon request.

SANTOS-DUMONT SMALL MODEL — REF. WSSA0023

Movement: quartz; approx. 6-year power reserve.
Functions: hours, minutes.
Case: steel; 38x27.5mm, thickness: 7.3mm; beaded crown set with a blue synthetic sapphire cabochon; sapphire crystal; water resistant to 3 bar.
Dial: silvered; sunray finish; blued steel sword-shaped hands.
Strap: navy blue semi-matte alligator leather; steel ardillon buckle.
Price: available upon request.

CARTIER

PANTHÈRE DE CARTIER MINI — REF. HPI01325

Movement: quartz.
Functions: hours, minutes.
Case: 18K white gold; 25x21mm, thickness: 6mm; set with brilliant-cut diamonds; octagonal crown set with a brilliant-cut diamond; sapphire crystal; water resistant to 3 bar.
Dial: silvered; blued steel sword-shaped hands.
Bracelet: 18K white gold; set with brilliant-cut diamonds.
Price: available upon request.

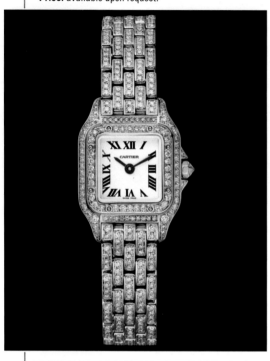

PANTHÈRE DE CARTIER MINI — REF. WJPN0020

Movement: quartz.
Functions: hours, minutes.
Case: 18K pink gold; 25x21mm, thickness: 6mm; octagonal crown set with a sapphire cabochon; sapphire crystal; water resistant to 3 bar.
Dial: silvered; blued steel sword-shaped hands.
Bracelet: 18K pink gold.
Price: available upon request.

PANTHÈRE DE CARTIER MINI — REF. WJPN0016

Movement: quartz.
Functions: hours, minutes.
Case: 18K yellow gold; 25x21mm, thickness: 6mm; octagonal crown set with a sapphire cabochon; sapphire crystal; water resistant to 3 bar.
Dial: silvered; blued steel sword-shaped hands.
Bracelet: 18K yellow gold.
Price: available upon request.

PANTHÈRE DE CARTIER MINI — REF. WSPN0019

Movement: quartz.
Functions: hours, minutes.
Case: steel; 25x21mm, thickness: 6mm; octagonal crown set with a synthetic spinel; sapphire crystal; water resistant to 3 bar.
Dial: silvered; blued steel sword-shaped hands.
Bracelet: steel.
Price: available upon request.

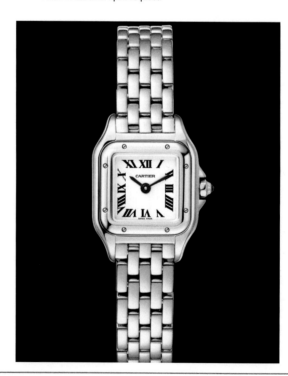

CARTIER

PANTHÈRE CRASH   REF. HPI01296

Movement: quartz.
Functions: hours, minutes.
Case: 18K white gold; Ø 46mmx23mm, thickness: 8.5mm; set with brilliant-cut diamonds, emerald eyes, black lacquer noses; sapphire crystal; water resistant to 3 bar.
Dial: mother-of-pearl; brilliant-cut diamond index; steel sword-shaped hands with rhodium finish.
Strap: black alligator leather; comes with second shiny pigeon blue alligator leather strap; 18K white-gold ardillon buckle set with brilliant-cut diamonds.
Price: available upon request.

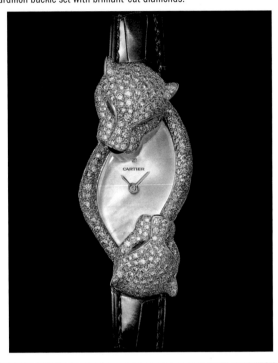

RONDE LOUIS CARTIER REGARD DE PANTHÈRE   REF. HPI01315

Movement: automatic-winding caliber 1847 MC; 40-hour power reserve.
Functions: hours, minutes.
Case: 18K yellow gold; Ø 36mm, thickness: 9.1mm; bezel set with brilliant-cut diamonds; yellow-gold beaded crown set with a brilliant-cut diamond; sapphire crystal; water resistant to 3 bar.
Dial: painted panther decor set with brilliant-cut diamonds and mother-of-pearl; eyes painted with SuperLumiNova; steel apple-shaped hands with yellow-gold finish.
Strap: shiny black alligator leather; comes with a second semi-matte brown alligator leather strap; 18K yellow-gold folding buckle set with brilliant-cut diamonds.
Note: limited edition of 30 individually numbered pieces.
Price: available upon request.

TANK CHINOISE   REF. WJLI0014

Movement: quartz.
Functions: hours, minutes.
Case: 18K white gold; 41x16mm, thickness: 7mm; set with brilliant-cut diamonds and rubies, black enamel; beaded crown set with a brilliant-cut diamond; sapphire crystal; water resistant to 3 bar.
Dial: black lacquer; steel sword-shaped hands.
Strap: black alligator leather; 18K white-gold ardillon buckle set with brilliant-cut diamonds.
Note: limited edition of 100 individually numbered pieces.
Price: available upon request.

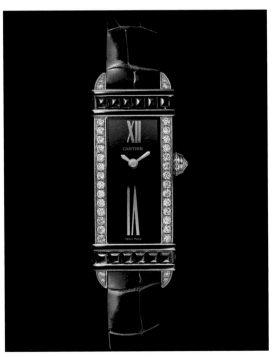

PANTHÈRE DE CARTIER   REF. HPI01297

Movement: quartz.
Functions: hours, minutes.
Case: 18K yellow gold; Ø 22mm, thickness: 7.7mm; bezel set with brilliant-cut diamonds, tsavorite eyes, black lacquer spots; sapphire crystal; water resistant to 3 bar.
Dial: golden; brilliant-cut diamond index; steel sword-shaped hands with yellow-gold finish.
Strap: gray alligator leather; 18K yellow-gold ardillon buckle.
Price: available upon request.

CHAUMET
PARIS

LOVE & LIGHT

With a history that spans over two centuries, Chaumet has earned its **EXPERTISE IN SOFT-FOCUS LUXURY THAT PERFECTLY EXPRESSES AN IDEA OR AN EMOTION**. Turning its attention to the delicate world of watches, the legendary jeweler uses a simple canvas to extend the concepts behind its flagship collections.

Chaumet's Hortensia Eden collection suggests a prelapsarian paradise, with the hydrangea blossoms picked out in precious gems and fine stones. The characteristic cluster of blooms pops up throughout the collection, each small flower retaining its individuality and charm. Crafted in 18-karat gold, pavé-set with round-cut diamonds, and cunningly carved from pieces of malachite and other stones, these flowers appeal to our need to capture these transient beauties in a more permanent medium. A series of watches complements the natural appeal of these jewels, pairing three individual blooms with a chic, minimalist dial. With unique slabs of malachite or a mirror-finished, rose-gold-tinted dial, the dial dispenses with hour markers or indexes, using only leaf-shaped hands to mark the time. The high-precision Swiss quartz movement inside can be set via a corrector on the caseback. Complementing the flower-shaped buckle, the three blossoms, crafted in 18-karat rose and gold and set with round-cut diamonds, each assert a subtle individuality in their construction.

▶ **HORTENSIA EDEN**
A selection of vivid straps complements the cheerful disposition of the rose-gold, diamond-set Hortensia Eden watch.

▶ **HORTENSIA EDEN**
(facing page)
Tiny blossoms signify the exuberant hydrangea in the Hortensia Eden collection, which includes jewelry as well as watches.

The characteristic cluster of blooms pops up throughout the Hortensia Eden collection, each small flower retaining its individuality and charm.

CHAUMET

The Liens family celebrates the joy of lasting love and the ties that bind. First introduced in 1970, the various aspects of familial, platonic, and romantic love find expression in playful designs that are each bound by a golden thread. Entering fully into the spirit of the collection, the Liens Lumière watch collection uses a distinctive case design to make a compelling reference to the gold ribbons that thread their way through the design family. Here they encircle the dial and connect with the elegant strap on which the watch is mounted. Even the ardillon buckle embraces the leitmotif of the collection, with crisscrossing straps that underscore the theme of connection and love. The second half of the "Liens Lumière" is evident as one admires the 339 brilliant-cut diamonds—a total of 2.44 carats—pavé-set into the 18-karat rose-gold bezel. A discreet date towards the bottom of the dial completes the elegant dial. The sapphire crystal caseback, also set with diamonds, reveals the automatic-winding Swiss mechanical movement within. A faceted diamond is set into the crown, completing the picture with a final kiss of luxury.

▲ **LIENS LUMIÈRE**
For an added touch of magic, some models in the Liens Lumière collection are adorned with diamond-set hour markers.

▶ **LIENS LUMIÈRE**
Pavé-set with 120 diamonds (0.14 carat), the caseback of the Liens Lumière presents an automatic movement with varied and sophisticated finishing.

▶ Encompassing watches and jewelry, the Liens collection uses a gold ribbon motif to signify lasting love. (*facing page*)

Chopard

BEST IN SHOW

Exploring the concept of excellence in a variety of fields, Chopard **MAKES REFERENCES TO VITICULTURE, CINEMA, AND THE HISTORY OF COMPLICATIONS** with three exceptional new models.

The shapes of watch cases may vary—circular, rectangular, oval—but the L.U.C Heritage Grand Cru stands out not just for the sophisticated "tonneau" shape of its case, but for the movement inside, which follows the same curvilinear lines. This commitment is singular in the world of watchmaking, which generally considers the shape of the movement as distinct from that of its housing. Crafted in white gold, the 38.5x38.8mm case of the L.U.C Heritage Grand Cru echoes the shape of a barrel used to age the finest of wines, adding a layer of significance to its name, which refers to the top oenological designation. An impeccably shaped outline of 40 baguette-cut diamonds limns the perimeter of the dial, bringing a bit of monochrome sparkle to this black and white piece. The self-winding L.U.C 97.01-L movement that powers the L.U.C Heritage Grand Cru boasts haute horology finishing and a diminutive thickness of just 3.30mm, allowing for a case slimness of 7.74mm. An open caseback reveals the wonder of the finely crafted tonneau-shaped L.U.C 97.01-L movement, which provides a power reserve of approximately 65 hours and boasts a 22-karat gold micro-rotor.

◀ **L.U.C HERITAGE GRAND CRU**
A modern tribute to Louis-Ulysse Chopard's pocket watches of the 1860s, this classic piece uses Roman numerals and railway-style minute track to convey the quiet elegance of watchmaking history, an endeavor crowned by the prestigious Poinçon de Genève.

CHOPARD

The L.U.C All-in-One, crafted entirely in the Chopard Haute Horlogerie workshops, incorporates the largest number of functions ever seen in an L.U.C watch.

The name of the L.U.C All-in-One is fitting, as this marvelous piece comprises 14 indications across two dials, folding immense complexity into a simple, elegant design. Available in two very limited editions—ten in platinum, ten in 18-karat rose gold—this grand complication watch, crafted entirely in the Chopard Haute Horlogerie workshops, incorporates the largest number of functions ever seen in an L.U.C watch. The recto side of the dial presents a serene view of the time, date, tourbillon, and perpetual calendar complete with 24-hour indication, date, day, month, and leap year. The breathtaking verso side reveals an equation of time display, power reserve, 24-hour day/night indication, sunrise and sunset times at Geneva, and astronomical orbital moonphase. The hand-wound L.U.C 05.01-L movement comprises 516 components, each one crafted and perfected at Chopard's own production sites.

Inspired by Chopard's longstanding partnership with the Cannes Film Festival, the Happy Palm borrows its central motif from the cinematic competition's most coveted prize: the Palme d'Or, created and crafted every year by the Swiss jeweler and watchmaker. In addition to the iconic mobile diamonds of the Happy Sport line, the Happy Palm adds a miniature golden palm that dances freely about the dial. The background is a fine example of the Japanese art of maki-e, using Urushi lacquer and gold dust to create an ethereal tableau of delicate fronds. The equally impressive caseback presents another alluring take on the theme, with more palm fronds engraved using the Fleurisanne technique, gently waving about the L.U.C 96.23-L movement. The Happy Palm model is emblematic in another way as well: it is crafted in 18-karat yellow Fairmined Gold, as part of Chopard's commitment to using only ethically sourced gold in all of its watches and jewelry.

▲ **L.U.C ALL-IN-ONE**
The 46mm case of the L.U.C All-in-One complements its solid gold guilloché dial, which is gray-blue on the platinum model, and an unorthodox verdigris for the rose-gold model.

▶ **HAPPY PALM**
This special edition, with five diamonds and a golden palm frond floating free about the dial, is released as part of the 25th anniversary of Chopard's Happy Sport collection, a development that definitively democratized daytime diamonds.

CHOPARD

L.U.C FULL STRIKE — REF. 161947-1001

Movement: manual-winding L.U.C 08.01-L caliber; Ø 37.2mm, thickness: 7.97mm; 60-hour power reserve; 533 components; 63 jewels; 28,800 vph; non-treated German silver mainplate and bridges; COSC-certified chronometer; Hallmark of Geneva.
Functions: hours, minutes; small seconds at 6; minute repeater.
Case: 18K "Fairmined" white gold; Ø 42.5mm, thickness: 11.55mm; antireflective sapphire crystal; hand-engraved caseback.
Dial: galvanic silver-toned gold; black rhodium-plated Roman numerals and hands.
Strap: black alligator leather; 18K "Fairmined" white-gold pin buckle.
Note: limited edition of 20 pieces.
Price: available upon request.

L.U.C ALL-IN-ONE — REF. 161925-5002

Movement: manual-winding L.U.C 05.01-L caliber; Ø 33mm, thickness: 11.75mm; 7-day power reserve; 516 components; 42 jewels; 28,800 vph; COSC-certified chronometer; Hallmark of Geneva.
Functions: hours, minutes; small seconds on tourbillon at 6; perpetual calendar: date at 12, day at 9, month and leap year at 3; 24-hour indicator at 9.
Case: 18K rose gold; Ø 46mm, thickness: 18.5mm; antireflective sapphire crystal; hand-engraved caseback; water resistant to 3atm.
Dial: hand-guilloché in a verdigris color obtained by galvanic treatment; gilded dauphine hands.
Strap: hand-stitched brown alligator leather; 18K rose-gold folding clasp.
Note: limited edition of 10 pieces.
Price: available upon request.
Also available: 950 platinum case, gray blue colored dial, blue alligator leather strap (ref. 161925-9003).

L.U.C QUATTRO — REF. 161926-5004

Movement: manual-winding L.U.C 98.01-L caliber; Ø 28.6mm, thickness: 3.7mm; 9-day power reserve; 223 components; 39 jewels; 28,800 vph; COSC-certified chronometer; Hallmark of Geneva.
Functions: hours, minutes; small seconds and date at 6; power reserve indicator at 12.
Case: 18K rose gold; Ø 43mm, thickness: 8.84mm; antireflective sapphire crystal; hand-engraved caseback; water resistant to 5atm.
Dial: vertical satin-brushed silver-toned; blued hour markers and hands.
Strap: blue calfskin leather; 18K rose-gold pin buckle.
Note: limited edition of 50 pieces.
Suggested price: $25,800

L.U.C HERITAGE GRAND CRU — REF. 172296-1001

Movement: automatic-winding L.U.C 97.01-L caliber; 28.15x27.6mm, thickness: 3.3mm; 65-hour power reserve; 197 components; 29 jewels; 28,800 vph; dual barrel; COSC-certified chronometer; Hallmark of Geneva.
Functions: hours, minutes; small seconds and date at 6.
Case: 18K white gold, 38.5x38.8mm, thickness: 7.74mm; bezel set with 40 baguette-cut diamonds (3.05 carats); antireflective sapphire crystal; exhibition caseback; water resistant to 3atm.
Dial: black lacquered; rhodium-plated dauphine hour and minute hands; rhodium-plated baton-shaped small seconds hand.
Strap: hand-stitched black alligator leather; 18K white-gold polished pin buckle.
Suggested price: $42,800

CHOPARD

L.U.C XPS 1860 RED CARPET EDITION — REF. 161946-1001

Movement: automatic-winding L.U.C 96.01-L caliber; Ø 27.4mm, thickness: 3.3mm; 65-hour power reserve; 29 jewels; 28,800 vph; dual barrel; COSC-certified chronometer; Hallmark of Geneva.
Functions: hours, minutes; small seconds and date at 6.
Case: 18K white gold; Ø 40mm, thickness: 7.2mm; antireflective sapphire crystal; exhibition caseback with Red Carpet logo; water resistant to 3atm.
Dial: red; hand-guilloché center, sunburst satin-finished exterior; rhodium-plated dauphine hands.
Strap: hand-stitched matte black alligator leather; 18K white-gold pin buckle.
Note: limited edition of 10 pieces.
Suggested price: $39,000

MILLE MIGLIA 2018 RACE EDITION — REF. 168589-3006

Movement: automatic-winding; Ø 28.6mm, thickness: 6.1mm; 42-hour power reserve; 37 jewels; 28,800 vph; COSC-certified chronometer.
Functions: hours, minutes, seconds; date between 4 and 5; chronograph: 12-hour counter at 6, 30-minute counter at 9, 60-second counter at 3.
Case: steel; Ø 42mm, thickness: 12.67mm; stainless steel bezel; antireflective sapphire crystal; exhibition back with special inscription "Chopard & Mille Miglia 30 anni di passione"; water resistant to 5atm.
Dial: engine-turned anthracite; gilded facetted hands; metallic black Arabic numerals with SuperLumiNova.
Strap: black calfskin leather; rubber lining inspired by 1960s Dunlop racing tire treads.
Note: limited edition of 1,000 pieces.
Suggested price: $5,620
Also available: steel and 18K rose-gold case, limited edition of 100 pieces (ref. 168589-6001).

MILLE MIGLIA GTS GRIGIO SPEZIALE — REF. 168566-3007

Movement: automatic-winding Chopard 01.08-C caliber; Ø 28.6mm, thickness: 4.95mm; 60-hour power reserve; 251 components; 40 jewels; 28,800 vph; COSC-certified chronometer.
Functions: hours, minutes, seconds; date at 3; power reserve indicator at 9.
Case: titanium; Ø 43mm, thickness: 11.43mm; DLC-treated blackened steel screw-down crown; antireflective sapphire crystal; exhibition caseback; water resistant to 10atm.
Dial: stippled anthracite gray ruthenium; matte black hour markers and Arabic numerals with SuperLumiNova.
Strap: Cordura fabric; steel folding clasp.
Note: limited edition of 1,000 pieces.
Suggested price: $8,220

MILLE MIGLIA RACING COLOURS — REF. 168589-3010

Movement: automatic-winding; Ø 28.6mm, thickness: 6.1mm; 42-hour power reserve; 37 jewels; 28,800 vph; COSC-certified chronometer.
Functions: hours, minutes, seconds; date between 4 and 5; chronograph: 12-hour counter at 6, 30-minute counter at 9, 60-second counter at 3.
Case: steel; Ø 42mm, thickness: 12.67mm; antireflective sapphire crystal; exhibition caseback; water resistant to 5atm.
Dial: satin-brushed Vintage Blue for France; dial bearing the traditional racing colors of the various nations competing in motorsports; Arabic numerals painted with SuperLumiNova; rhodium-plated baton-type hands with SuperLumiNova.
Strap: calfskin leather; rubber lining inspired by 1960s Dunlop tire treads; steel pin buckle.
Note: limited edition of 300 pieces; boutique and corner exclusive.
Suggested price: $6,080
Also available: Rossa Corsa dial for Italy (ref. 168589-3008); Speed Silver dial for Germany (ref. 168589-3012); British Racing Green dial for the United Kingdom (ref. 168589-3009); Speed Yellow dial for Belgium (ref. 168589-3011).

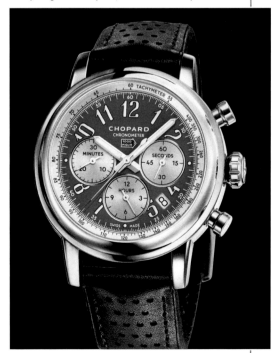

CHOPARD

GRAND PRIX DE MONACO HISTORIQUE 2018 RACE EDITION   REF. 168570-9002

Movement: automatic-winding; Ø 37.2mm; 46-hour power reserve; 25 jewels; 28,800 vph; COSC-certified chronometer.
Functions: hours, minutes; seconds; date at 3; chronograph: 12-hour counter at 6, 30-minute counter at 12, 60-second counter at 9.
Case: shotblasted titanium and steel; Ø 44.5mm, thickness: 14.1mm; antireflective sapphire crystal; caseback stamped with the Automobile Club de Monaco; water resistant to 10atm.
Dial: silver-toned; galvanic blue counters at 12 and 6; rhodium-coated hour and minute hands with Super-LumiNova.
Strap: blue Barenia calfskin leather with orange stitching; blue and orange NATO strap; titanium and steel folding clasp.
Note: limited edition of 250 pieces.
Suggested price: $7,390
Also available: 18K rose-gold, titanium and steel case, limited edition of 100 pieces (ref. 168570-3004).

L.U.C XP ESPRIT DE FLEURIER PEONY   REF. 131944-5003

Movement: automatic-winding L.U.C 96.23-L caliber; Ø 27.4mm, thickness: 3.5mm; 65-hour power reserve; 168 components; 29 jewels; 28,800 vph; dual barrel.
Case: 18K rose gold; Ø 35mm, thickness: 7.7mm; diamond-set bezel, sides and lugs; exhibition back with antireflective sapphire crystal; water resistant to 3atm.
Dial: gold coated with Grand Feu enamel featuring a black peony motif crafted using the paper-cutting technique typical of the Pays d'Enhaut; gilded dauphine hour and minute hands.
Strap: brushed black canvas; diamond-set polished 18K rose-gold pin buckle.
Note: limited edition of 8 pieces.
Price: available upon request.

HAPPY SPORT JOAILLERIE   REF. 274891-1016

Movement: automatic-winding; Ø 26.2mm, thickness: 3.6mm; 42-hour power reserve; 25 jewels; 28,800 vph.
Functions: hours, minutes, seconds.
Case: 18K white gold; Ø 36mm, thickness: 12.65mm; snow-set bezel; 18K white-gold facetted crown with a diamond; antireflective sapphire crystal; exhibition caseback; water resistant to 3atm.
Dial: textured blue mother-of-pearl; seven moving diamonds; rhodium-plated hour markers set with diamonds apart from the four Roman numerals; rhodium-plated hands.
Strap: royal blue alligator leather; 18K white-gold pin buckle.
Note: limited edition of 25 pieces.
Suggested price: $66,900

HAPPY FISH   REF. 274891-5019

Movement: automatic-winding; Ø 26.2mm, thickness: 3.6mm; 42-hour power reserve; 25 jewels; 28,800 vph.
Functions: hours, minutes, seconds.
Case: 18K rose gold; Ø 36mm, thickness: 12.65mm; bezel set with diamonds of three different sizes; 18K rose-gold facetted crown set with a diamond; antireflective sapphire crystal; exhibition caseback bearing the Happy Fish inscription; water resistant to 3atm.
Dial: gold; snow-set with subtly graded shades of dark blue sapphires with domed fish in textured mother-of-pearl; seven luminescent moving diamonds; gilded hands.
Strap: royal blue alligator leather; 18K rose-gold pin buckle.
Note: limited edition of 25 pieces; boutique exclusive.
Suggested price: $52,300

CHOPARD

HAPPY SPORT — REF. 274808-5015

Movement: automatic-winding; Ø 26.2mm, thickness: 3.6mm; 42-hour power reserve; 25 jewels; 28,800 vph.
Functions: hours, minutes, seconds; date between 4 and 5.
Case: 18K rose gold; Ø 36mm; diamond-set bezel;
Dial: gray guilloché; seven moving diamonds.
Bracelet: 18K rose gold.
Suggested price: $39,400

HAPPY SPORT — REF. 278573-3010

Movement: automatic-winding Chopard 09.01-C caliber; Ø 20.4mm, thickness: 3.65mm; 42-hour power reserve; 159 components; 27 jewels; 25,200 vph.
Functions: hours, minutes, seconds.
Case: steel; Ø 30mm; thickness: 10.95mm; diamond-set steel bezel; facetted steel crown set with a blue sapphire; antireflective sapphire crystal; exhibition caseback with Happy Sport logo; water resistant to 3atm.
Dial: light blue textured mother-of-pearl; five moving diamonds; rhodium-plated Roman numerals and hour markers; rhodium-plated hands.
Strap: lavender blue alligator leather; polished steel pin buckle.
Suggested price: $14,200
Also available: steel and 18K rose-gold case, bezel without diamonds, pink mother-of-pearl dial, matte pink alligator leather strap (ref. 278573-6011); 18K rose-gold case, bezel without diamonds, white mother-of-pearl dial, matte white alligator leather strap (ref. 274893-5009); 18K rose-gold case, white mother-of-pearl dial, matte white alligator leather (ref. 274893-5010).

HAPPY SPORT OVAL — REF. 275362-5003

Movement: automatic-winding Chopard 09.01-C caliber; Ø 20.4mm, thickness: 3.65mm; 42-hour power reserve; 159 components; 27 jewels; 25,200 vph.
Functions: hours, minutes, seconds.
Case: 18K rose gold; 31.31x29mm, thickness: 10.77mm; 18K rose-gold bezel set with pink sapphires and rubies; facetted 18K rose-gold crown set with a blue sapphire; antireflective sapphire crystal; exhibition caseback with Happy Sport logo; water resistant to 3atm.
Dial: white mother-of-pearl with silver guilloché center; two moving diamonds, two moving pink sapphires, three moving rubies; gilded Roman numerals and hour markers; gilded hands.
Strap: fuchsia alligator leather; 18K rose-gold pin buckle.
Suggested price: $19,400
Also available: silver guilloché dial, royal blue alligator leather strap (ref. 275362-5001); bezel set with diamonds, navy blue alligator leather strap (ref. 275362-5002).

L'HEURE DU DIAMANT — REF. 10A375-1001

Movement: quartz.
Functions: hours, minutes.
Case: 18K white gold; 39.2x33m; set with round brilliant-cut diamonds (6.93 carats).
Dial: opal.
Bracelet: 18K white gold; set with round brilliant-cut diamonds (23.44 carats).
Price: available upon request.

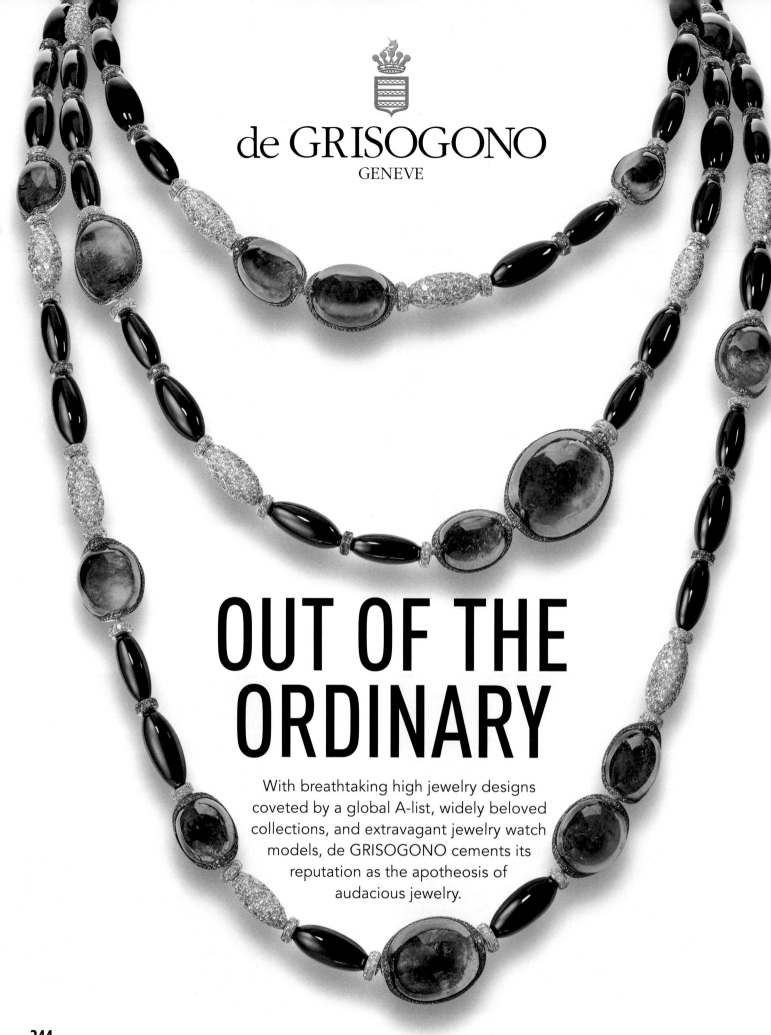

de GRISOGONO
GENEVE

OUT OF THE ORDINARY

With breathtaking high jewelry designs coveted by a global A-list, widely beloved collections, and extravagant jewelry watch models, de GRISOGONO cements its reputation as the apotheosis of audacious jewelry.

de GRISOGONO

de GRISOGONO's high jewelry collections make headlines and dominate red carpets at the most prestigious and glamorous events across the globe. The world's most beautiful and talented women vie to wear the Swiss jeweler's inventive, irreverent designs on an international stage, at events such as the renowned Cannes Film Festival. When all eyes are on them, women adore being admired in the gorgeous—and often outrageous—designs that emanate from the jeweler's Geneva atelier. In the most elegant and dynamic settings, de GRISOGONO high jewelry makes a statement, sometimes stealing the show in the process.

▶ Amber Heard wears de GRISOGONO High Jewelry ruby earrings at the Cannes Film Festival. Crafted in 18-karat white gold and 18-karat pink gold, the earrings are set with 1 pear-cut white diamond E VS (1.00 carat), 1 pear-cut white diamond F VVS2 (1.06 carat), 1 pear-cut ruby Vivid Red no heat (7.54 carats), 1 pear-cut ruby Vivid Red no heat (7.57 carats), 2 trapeze-cut rubies (0.71 carats), 48 baguette-cut white diamonds (6.52 carats), 164 rubies (0.50 carat), and 366 white diamonds (0.52 carat).

◀ Alessandra Ambrosio wears a de GRISOGONO High Jewelry necklace at the Cannes Film Festival, crafted in 18-karat white gold and set with 1 cabochon-cut emerald (43.64 carats), 14 cabochon-cut emeralds (157.41 carats), 65 onyx (288.07 carats), 1,416 white diamonds (23.57 carats), and 1,349 emeralds (7.50 carats), as well as a pair of earrings from the High Jewelry collection in 18-karat white gold set with 6 emeralds (68.89 carats), 137 emeralds (0.46 carat), and 677 diamonds (7.75 carats).

de GRISOGONO

Three things set de GRISOGONO apart from other makers of high jewelry. The jeweler uses only the highest-quality precious stones, with diamonds acquired directly from some of the world's finest mines. Exceptional rubies, sapphires, and emeralds bring rich hues and a nuanced gleam to pieces conceived to highlight their beauty. The workmanship throughout every collection is absolutely exquisite, with a team of master gem-setters, polishers, and goldsmiths devoted to realizing every piece in its fully individualized glory. At the very foundation of de GRISOGONO's success is the constant stream of daring, bold designs from which its pieces grow into reality.

◀ Romee Strijd wears a de GRISOGONO High Jewelry necklace at the Cannes Film Festival, crafted in 18-karat white and set with 1 pear-cut white diamond (42.53 carats), 1 pear-cut white diamond (0.19 carat), 1,112 white diamonds (57.62 carats), and 1,570 blue sapphires (125.50 carats).

de GRISOGONO

Some designs are conceived around a particular stone—an enormous pear-shaped diamond becomes the seed—and the centerpiece—of a stunning necklace, pavé-set with precious stones in a thick impasto of sparkle that gradually changes from the pure white of diamonds to a rich, deep blue of sapphires. The necklace practically demands a plunging neckline to match its drama and dynamism. Startlingly blue sapphires also make an appearance on a collar necklace that presents the generous stones as if randomly placed on a bed of diamonds. This piece epitomizes the de GRISOGONO approach in a myriad of ways, from the flawless blue sapphires front and center, to the hidden sapphire setting at the collar's top, to the equally hidden volutes engraved freehand in the interior of the 18-karat white-gold piece: a signature of the visionary jeweler. The diamond setting is also characteristic of de GRISOGONO: the snow-setting technique, in the way it develops organically by sitting individual stones one at a time, seems to make of the necklace a work of nature, like the fascinating frost crystals that colonize a glass window in a snowy Alpine winter. The jeweler's trademark Icy diamonds, a semi-opaque take on the precious stone, gleam with a soft light, visually underscoring the sapphires' clear blue all the more.

◀ Cindy Bruna wears a de GRISOGONO High Jewelry necklace at the Cannes Film Festival, crafted in 18-karat white gold and set with 1 blue cushion-cut sapphire (4.68 carats), 1 blue cushion-cut sapphire (5.70 carats), 1 blue cushion-cut sapphire (5.86 carats), 1 blue cushion-cut sapphire (6.29 carats), 1 blue cushion-cut sapphire (6.43 carats), 1 blue cushion-cut sapphire (8.18 carats), 1 pear-cut blue sapphire (0.53 carat), 170 blue sapphires (7.36 carats), 477 icy diamonds (56.81 carats), and 1,104 white diamonds (28.7 carats).

de GRISOGONO

▶ **ALLEGRA TOI & MOI BRACELET**
Crafted in 18-karat white gold and set with 126 white diamonds (approx. 0.75 carat).

▶ **ALLEGRA CLASSIC RING**
Crafted in 18-karat pink gold and set with 100 white diamonds (approx. 2.95 carats).

Never content to rest on its laurels, de GRISOGONO continues to develop its most popular collections as well as new ones, adding twists and turns that continue to define the unique vision of the Swiss jeweler. The Classic Allegra collection expresses the essence of joy, with diamond-set hoops that dance around the piece. Whether as a ring, bracelet, earrings, or necklace, Allegra enhances every occasion, with Toi & Moi variations that put a modern spin on the ultra-romantic jewelry concept. Ventaglio, inspired by the supreme elegance of the fan, captures the vintage charm of the Roaring Twenties. Another collection harkening back to the Jazz Age, Raggiante, whose name means "radiant," also brings to mind the exuberant energy of ragtime, with diamond-set bursts radiating out from the band. Divina draws from the restrained genius of the 1950s and such icons as Maria Callas and Jacqueline Kennedy, its very silhouette following the nipped waist and full, flowing skirt that dominated the decade's fashion. A liberated energy flows through Tentazione, a spiritual visit to the hedonistic extravagance of the Studio 54 era. The thread that binds all these ideas together is a masterful use of gold and diamonds, as well as a certain inimitable insouciance.

▲ **ALLEGRA TOI & MOI EARRINGS**
Crafted in 18-karat rose gold and set with white diamonds and 2 cacholong cabochons.

▲ **ALLEGRA TOI & MOI EARRINGS**
Crafted in 18-karat white gold and set with 248 white diamonds (approx. 0.75 carat) and 4 turquoise cabochons (approx. 0.9 carat).

de GRISOGONO

◀ EXTRAVAGANZA DIVINA CUFF
Crafted in 18-karat pink gold and set with 287 white diamonds (approx. 27.00 carats).

▼ EXTRAVAGANZA DIVINA EARRINGS
Crafted in 18-karat pink gold and set with 478 white diamonds (approx. 14.20 carats).

▲ VENTAGLIO EARRINGS
Crafted in 18-karat white gold and set with 230 white diamonds (8.65 carats) and 240 icy diamonds (approx. 9.65 carats).

▲ EXTRAVAGANZA TENTAZIONE EARRINGS
Crafted in 18-karat white gold and set with 444 white diamonds (approx. 7 carats).

CASCATA

With a name that evokes a sparkling waterfall, Cascata evokes the fleeting beauty of running water, and it takes its commitment to luxury very seriously. Thick rows of oval-cut diamonds surround a snow-set dial, the diamonds' round shapes and fortuitous placement seeming to costume the timepiece in the guise of a bubbling natural pond. Brimming with vitality, the jewelry watch shows off the expertise of its creation from every angle. The exquisite bezel bears 72 oval-cut diamonds, held in place by tiny prongs, which are themselves paved with 126 brilliant-cut diamonds. The snow-set dial scintillates with 148 diamonds itself. Even the caseback is a thing of beauty, outlining de GRISOGONO's trademark volutes in diamonds.

▼ **CASCATA**
Crafted in 18-karat white gold, the 48x40mm Cascata watch is set with 148 diamonds on the dial, 72 oval-cut diamonds and 254 brilliant-cut diamonds on the case, and 95 diamonds on the folding clasp which fastens the white galuchat strap.

ALLEGRA 25

Celebrating two and a half decades of de GRISOGONO's jewelry vision, the Allegra 25 watch embodies the energy of this youthful Maison. Black diamonds set on the dial and case nod to the Swiss jeweler's role as the first champion of this intriguing stone. An 18-karat white-gold ribbon dances at the edge of the dial, flowing around the periphery from 3 to 9 o'clock, with 29 gem-set rings executing the collection's beloved swoops. Building on the contrast between white and black diamonds, the 41.7mm 18-karat white-gold case is mounted on a galuchat strap—another foundational element of de GRISOGONO's design identity.

▶ **ALLEGRA 25**
Set with 555 black diamonds and 103 white diamonds, and mounted on a galuchat strap, the celebratory Allegra 25 uses several de GRISOGONO signature motifs to celebrate the jeweler's rich history.

de GRISOGONO

NEW RETRO

Though its clean lines hearken back to the design codes of the 1950s, de GRISOGONO's New Retro watch has always been pure 21st century. With a timeless design easily adaptable to masculine or feminine sensibilities, the New Retro won hearts the world over. The addition of baguette-cut diamonds only enhances its unmistakable allure. The New Retro Lady Baguette is graced with a wall-to-wall impasto of glitter, with case, dial, and bracelet entirely covered with baguette-cut diamonds. The rectangular shape of the stones emphasizes anew the geometric appeal of the New Retro design, an instant icon ever since its release.

▶ **NEW RETRO**
Unmistakably bold, the New Retro Lady Baguette in 18-karat rose gold adds a dense layer of baguette-cut diamonds to the versatile design—1,151 diamonds total (36.85 carats).

ECCENTRICA

The aptly named Eccentrica also explores the possibilities of baguette-cut diamonds, taking the concept, as de GRISOGONO so often does, to the extreme. The dial, case, lugs, crown, and even bracelet of this model are all completely covered in white diamonds, the vast majority baguette-cut. The dial's concentric circles suggest a strong, self-confident aesthetic vision, with an underlying feeling of harmony, as well as a dizzying sense of depth. The mechanical self-winding movement within, DG 10-01, is the watch lover's cherry on top of this extravagant jewelry aficionado's sundae.

◀ **ECCENTRICA**
A little off-kilter, but gorgeous nonetheless, the Eccentrica watch boasts no end of sparkle. The watch is crafted in 18-karat white gold and set with 1,586 diamonds (42.62 carats).

DIOR

INVITATION TO THE DANCE

A PASSION FOR FLOWERS AND CONSUMMATE HIGH FASHION MEETS ADVENTUROUS HOROLOGY in the Dior Grand Bal Miss Dior, available in two new limited edition models. The entrancing dial uses unconventional techniques to create a wholly new aesthetic effect, bringing a touch of fantastical luxury to the natural world.

DIOR

The automatic Dior Inversé 11 ½ caliber of the Dior Grand Bal Miss Dior has been constructed "upside-down," turning the weight into an aesthetic element.

The watches draw their shared name in part from the "Miss Dior" dress, one of Christian Dior's first haute couture dresses, crafted in 1949. The dress was a floral vision of femininity, covered in hand-sewn blossoms and flaring out from the waist like a flower itself. The Dior Grand Bal Miss Dior takes this vision into the 21st century, reinterpreting the floral theme on the dial.

Two designs illustrate the duality of the concept behind the design. A model in vibrant shades of pink evokes the striking vision of intense summer sunlight on gaily colored petals, almost intoxicating in the vital promise it offers. A sultry blue model takes us under the moonlight, with a black periphery and cool hues that suggest a midnight garden stroll illuminated only by "l'astre de nuit."

In an unusual configuration, the movement's oscillating weight takes pride of place on the dial, performing its crucial function front and center, not hidden away as it so often is. The dial itself, displaying the hours and minutes with pink-gold hands, is crafted in mother-of-pearl, engraved with a sunray pattern and boldly hued with a tie-dyed shaded effect: an extraordinary base for what rests upon it. The automatic Dior Inversé 11 ½ caliber, as its name suggests, has been constructed "upside-down," turning the weight into an aesthetic element. The openworked pink-gold weight has been adorned with translucent ceramic petals, depicting a three-dimensional bloom on the dial that seems to move as if blown by the wind. Lacquered pink-gold corollas and round-cut diamonds add an eerie sense of magic to this mobile element, which provides a power reserve of 42 hours.

Diamonds also adorn each 36mm case, in a snow setting that underscores the allure of natural chance, as the stones are set in place one at a time, following no prearranged pattern. The case is fashioned in steel, with a pink-gold bezel and crown to lend variety and frame the precious flower created by the dial. Water resistant to 50m, the case sports a sapphire crystal caseback in colors to match each model, further decorated with pink-gold-colored metallization. The black leather strap bearing this extraordinary piece sports a diamond-set stainless steel prong buckle, as well as a lining to match the colors of the watch—a secret known only to its wearer. This stunning flower for the wrist is available in a limited edition of 88 pieces for each version.

◀ **DIOR GRAND BAL MISS DIOR**
Set with a total of 392 diamonds (2.32 carats), the Dior Grand Bal Miss Dior brings a kiss of luxury to the natural beauty of a flower.

▲ **DIOR GRAND BAL MISS DIOR**
The Dior Inversé 11 ½ movement reveals its oscillating weight on the dial, and the rest of its workings through the sapphire crystal caseback.

DIOR

LA MINI D DE DIOR SATINE — REF. CD040120M001

Movement: quartz.
Functions: hours, minutes.
Case: stainless steel; Ø 19mm; 18K yellow-gold bezel set with diamonds; 18K yellow-gold crown set with diamonds; water resistant to 3atm.
Dial: malachite.
Bracelet: Milanese mesh stainless steel; unfolding steel buckle set with diamonds.
Note: set with 73 diamonds (approx. 0.47 ct).
Price: available upon request.

LA MINI D DE DIOR SATINE — REF. CD040120M002

Movement: quartz.
Functions: hours, minutes.
Case: stainless steel; Ø 19mm; 18K yellow-gold bezel set with diamonds; 18K yellow-gold crown set with diamonds; water resistant to 3atm.
Dial: lapis lazuli.
Bracelet: Milanese mesh stainless steel; unfolding steel buckle set with diamonds.
Note: set with 73 diamonds (approx. 0.47 ct).
Price: available upon request.

LA MINI D DE DIOR SATINE — REF. CD040110M001

Movement: quartz.
Functions: hours, minutes.
Case: stainless steel; Ø 19mm; bezel set with diamonds; crown set with diamonds; water resistant to 3atm.
Dial: white mother-of-pearl.
Bracelet: Milanese mesh stainless steel; unfolding steel buckle set with diamonds.
Note: set with 73 diamonds (approx. 0.47 ct).
Price: available upon request.

LA MINI D DE DIOR SATINE TRESSEE — REF. CD040150M002

Movement: quartz.
Functions: hours, minutes.
Case: 18K yellow gold; Ø 19mm; 18K yellow-gold bezel set with diamonds; 18K yellow-gold crown set with diamonds; water resistant to 3atm.
Dial: golden sun brushing.
Strap: polished and chased Polish mesh in yellow gold; yellow-gold unfolding buckle set with diamonds.
Note: set with 73 diamonds (approx. 0.47 ct).
Price: available upon request.

DIOR

LA D DE DIOR SATINE — REF. CD047111M001

Movement: quartz.
Functions: hours, minutes.
Case: stainless steel; Ø 25mm; bezel set with diamonds; crown set with diamonds; water resistant to 3atm.
Dial: white mother-of-pearl; set with 4 diamonds.
Bracelet: Milanese mesh stainless steel; unfolding steel buckle set with diamonds.
Note: set with 91 diamonds (approx. 0.70 ct).
Price: available upon request.

LA D DE DIOR SATINE — REF. CD047111M002

Movement: quartz.
Functions: hours, minutes.
Case: stainless steel; Ø 25mm; bezel set with diamonds; crown set with diamonds; water resistant to 3atm.
Dial: black mother-of-pearl; set with 4 diamonds.
Bracelet: Milanese mesh stainless steel; unfolding steel buckle set with diamonds.
Note: set with 91 diamonds (approx. 0.70 ct).
Price: available upon request.

LA D DE DIOR SATINE — REF. CD043120M001

Movement: quartz.
Functions: hours, minutes.
Case: stainless steel; Ø 36mm; 18K yellow-gold bezel set with diamonds; 18K yellow-gold crown set with diamonds; water resistant to 3atm.
Dial: white mother-of-pearl; set with 12 diamonds.
Bracelet: Milanese mesh stainless steel; unfolding steel buckle set with diamonds.
Note: set with 128 diamonds (approx. 1.08 ct).
Price: available upon request.

LA D DE DIOR SATINE — REF. CD043120M002

Movement: quartz.
Functions: hours, minutes.
Case: stainless steel; Ø 36mm; 18K yellow-gold bezel set with diamonds; 18K yellow-gold crown set with diamonds; water resistant to 3atm.
Dial: malachite.
Bracelet: Milanese mesh stainless steel; unfolding steel buckle set with diamonds.
Note: set with 116 diamonds (approx. 0.98 ct).
Price: available upon request.

DIOR

LA D DE DIOR — REF. CD043114A001

Movement: quartz.
Functions: hours, minutes.
Case: stainless steel; Ø 38mm; bezel set with diamonds; crown set with diamonds; water resistant to 3atm.
Dial: white mother-of-pearl; set with 12 diamonds.
Strap: black satin; steel prong buckle set with diamonds.
Note: set with 120 diamonds (approx. 1.14 ct).
Price: available upon request.

LA D DE DIOR — REF. CD043171A001

Movement: quartz.
Functions: hours, minutes.
Case: 18K pink gold; Ø 38mm; 18K pink-gold bezel set with diamonds; 18K pink-gold crown set with diamonds; water resistant to 3atm.
Dial: white mother-of-pearl; set with 12 diamonds.
Strap: black satin; 18K pink-gold prong buckle set with diamonds.
Note: set with 156 diamonds (approx. 1.29 ct).
Price: available upon request.

LA D DE DIOR — REF. CD043171A002

Movement: quartz.
Functions: hours, minutes.
Case: 18K pink gold; Ø 38mm; 18K pink-gold bezel set with diamonds; 18K pink-gold crown set with diamonds; water resistant to 3atm.
Dial: onyx; set with 12 diamonds.
Strap: black satin; 18K pink-gold prong buckle set with diamonds.
Note: set with 156 diamonds (approx. 1.29 ct).
Price: available upon request.

LA D DE DIOR — REF. CD043171A003

Movement: quartz.
Functions: hours, minutes.
Case: 18K pink gold; Ø 38mm; 18K pink-gold bezel set with diamonds; 18K pink-gold crown set with diamonds; water resistant to 3atm.
Dial: jadeite jade set with 12 diamonds.
Strap: black satin; 18K pink-gold prong buckle set with diamonds.
Note: set with 156 diamonds (approx. 1.29 ct).
Price: available upon request.

DIOR

DIOR GRAND BAL RESILLE ROUGE — REF. CD153B14A001

Movement: automatic; "Dior inverse 11 ½" caliber; functional white-gold oscillating weight on the dial; set with 182 round-cut diamonds (patented); 42-hour power reserve.
Functions: hours, minutes.
Case: polished stainless steel; Ø 36mm; stainless steel bezel with a mother-of-pearl ring; antireflective sapphire crystal glass; stainless steel crown engraved with "CD"; translucent red sapphire crystal caseback; water resistant to 5atm.
Dial: red sun-brushed.
Strap: shiny red alligator leather; steel prong buckle.
Note: limited edition of 88 pieces; set with 182 diamonds (approx. 0.33 ct).
Price: available upon request.

DIOR GRAND BAL PLUME — REF. CD153B2GA001

Movement: automatic; "Dior inverse 11 ½" caliber; functional pink-gold oscillating weight on the dial; decorated with white feathers and set with round-cut diamonds (patented); 42-hour power reserve.
Functions: hours, minutes.
Case: polished stainless steel; Ø 36mm; stainless steel bezel set with round-cut diamonds and decorated with a pink-gold ring; antireflective sapphire crystal glass; stainless steel crown engraved with "CD"; translucent sapphire crystal caseback with golden metalization and shaded blue effect; water resistant to 5atm.
Dial: aventurine.
Strap: shiny blue alligator leather; steel prong buckle set with round-cut diamonds.
Note: limited edition of 88 pieces; set with 220 diamonds (approx. 1.06 ct).
Price: available upon request.

DIOR GRAND BAL PLUME — REF. CD153B2FA001

Movement: automatic; "Dior inverse 11 ½" caliber; functional yellow-gold oscillating weight on the dial; decorated with white feathers and set with round-cut diamonds (patented); 42-hour power reserve.
Functions: hours, minutes.
Case: polished stainless steel; Ø 36mm; stainless steel bezel set with round-cut diamonds and decorated with a yellow-gold ring; antireflective sapphire crystal glass; stainless steel crown engraved with "CD"; translucent sapphire crystal caseback with golden metalization and shaded green effect; water resistant to 5atm.
Dial: malachite.
Strap: shiny green alligator leather; steel prong buckle set with round-cut diamonds.
Note: limited edition of 88 pieces; set with 220 diamonds (approx. 1.06 ct).
Price: available upon request.

DIOR GRAND BAL PLUME NOIRE — REF. CD153B2BA001

Movement: automatic; "Dior inverse 11 ½" caliber; functional yellow-gold oscillating weight on the dial; decorated with black feathers and set with round-cut diamonds; 42-hour power reserve.
Functions: hours, minutes.
Case: polished stainless steel; Ø 36mm; stainless steel bezel set with round-cut diamonds and decorated with a yellow-gold ring; antireflective sapphire crystal glass; stainless steel crown engraved with "CD"; yellow-gold metalized sapphire crystal caseback; water resistant to 5atm.
Dial: black sun-brushed; dial set with yellow gold threads.
Strap: semi-matte black alligator leather; steel prong buckle set with round-cut diamonds.
Note: limited edition of 88 pieces; set with 197 diamonds (approx. 1.03 ct).
Price: available upon request.

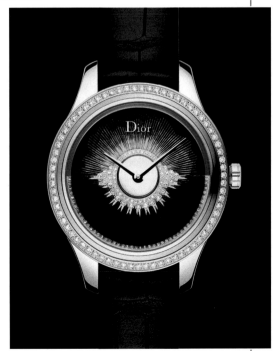

FRANCK MULLER
GENEVE

INAUGURATION AND INNOVATION

KNOWN FOR ITS GROUNDBREAKING, EYE-CATCHING DESIGN, luxury watchmaking maison Franck Muller recently unveiled two new buildings added to its unique Switzerland-based Watchland headquarters, along with its legendary and no less complex Crazy Hours complication, this time dedicated to female wrists.

The Watchland extension, located in Genthod, consists of two new buildings that first began to emerge in 2018. They represent the perfect finishing touch to the Geneva location and enable the group to unite its 450 employees and all stages in the production of Franck Muller watches together on one site. The spectacular Genthod estate has been reinvented and expanded to accommodate 16,000m^2 of additional working space on the Watchland site. All the artisans required at each stage of watch production are now together under one roof, demonstrating once and for all the fantastic success enjoyed by the brand, and its exceptional expertise in the field of grand complications. Located at the heart of Swiss watchmaking, the property centralizes the company's activities and now houses the watch component manufacturing team previously based in Satigny. The architecture is subtly reminiscent of Versailles, with the two new buildings accentuating the distinctive character of the Watchland domain. The classic terraced gardens of the premises, along with the breathtaking views of Lake Geneva and Mont Blanc, are just some of the features that make this location absolutely unique. Faithful to his roots, Franck Muller chose Genthod for the headquarters because he began his activity in this village in 1983.

The brand that has become, in less than 30 years, a benchmark in the area of haute horology, also recently introduced the Vanguard collection, to which Franck Muller added its famed complication the Crazy Hours, with a feminine twist. Bold, startling, and expertly crafted, this timepiece displays numerals in an unconventional order, with an atypical Art Deco, Pop Art style. Driven by a patented mechanism, the central hours hand jumps from one hour to the next, following the correct sequence of numerals, even though they are randomly placed on the dial. Every 60 minutes, the central hand jumps to the next correct numeral, while the minutes hand continues its conventional sweep around the dial. This novel and unique interpretation of time becomes a fascinating visual game. Available in a variety of colors, the timepiece displays the jumping hours, minutes and seconds functions thanks to a self-winding mechanical movement, whose heart beats at 28,800 vph, and features a 40-hour power reserve. Enlivened by a movement structure composed of alternating concave and convex bridges, the 18-karat pink-gold case, teamed with an alligator leather strap, has an undeniably striking, iconic design. The movement decoration features diamond polishing and Côtes de Genève on the bridges, 24-karat gold finish on the engravings, 24-karat gold bath and rhodium treatment, and sunray brushing on the oscillating weight. The Vanguard Crazy Hours Lady is designed to express the concept of time as an abstract notion: that time is ultimately what one makes of it.

FRANCK MULLER

The novel and unique interpretation of time on the Vanguard Crazy Hours Lady model becomes a fascinating visual game.

◀ Bold, startling, and expertly crafted, the Crazy Hours timepiece displays numerals in an unconventional order, with an atypical Art Deco, Pop Art style.

◀ The stunning Switzerland-based Watchland domain now benefits from a two-building extension, reuniting all stages in the production of Franck Muller watches on one site.
(*facing page*)

FRANCK MULLER

8883 CC GD – CHRONOGRAPHE GRANDE DATE

Movement: automatic-winding chronograph.
Functions: hours, minutes; large date at 7; 2-counter chronograph.
Case: 5N pink gold; "Cintrée Curvex"; sapphire crystal; transparent caseback; water resistant to 30m.
Dial: white; guilloché; partially openworked; available in several colors.
Strap: alligator leather.
Also available: in white gold; in steel.

8880 MASTER BANKER

Movement: automatic-winding.
Functions: hours, minutes; date; three time zones adjustable to the minute, set via the crown.
Case: steel; "Cintrée Curvex"; sapphire crystal; water resistant to 30m.
Dial: white; guilloché; available in several colors.
Strap: alligator leather.
Also available: in white gold; in pink gold.

8880 C CC DT – CASABLANCA CHRONOGRAPHE

Movement: automatic-winding chronograph.
Functions: hours, minutes; date at 6; 2-counter chronograph.
Case: pink gold; "Cintrée Curvex"; sapphire crystal; water resistant to 30m.
Dial: white; sunray guilloché; available in several colors.
Strap: smooth calfskin.
Also available: in yellow gold; in white gold; in steel.

VEGAS 8880 CINTREE CURVEX

Movement: automatic-winding; oscillating weight in 950 platinum.
Functions: hours, minutes; independent roulette wheel activated by co-axial mono-pusher at 3, a Franck Muller exclusive.
Case: pink gold; "Cintrée Curvex"; sapphire crystal; water resistant to 30m.
Dial: sunray guilloché; available in several colors.
Strap: alligator leather.
Also available: in yellow gold; in white gold; in steel.

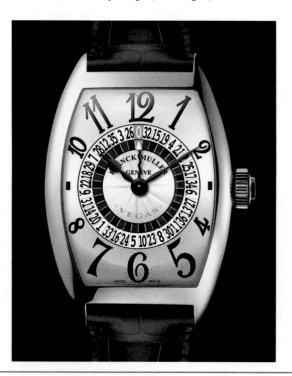

FRANCK MULLER

1752 QZ D COLOR DREAM D – LADY CINTREE CURVEX

Movement: quartz.
Functions: hours, minutes.
Case: pink gold; "Cintrée Curvex"; set with diamonds; sapphire crystal; water resistant to 30m.
Dial: white; sunray guilloché; "Color Dream" numerals; available in several colors.
Strap: alligator leather.
Also available: in yellow gold; in white gold; in steel.

DOUBLE MYSTERY D CD

Movement: automatic-winding "Double Mystery" caliber; oscillating weight in 950 platinum.
Functions: hours, minutes; "mysterious" time indication via two rotating disks, Franck Muller exclusive, patented by Franck Muller.
Case: white gold; Ø 42mm; fully set with diamonds; sapphire crystal; transparent caseback; water resistant to 30m.
Dial: two rotating disks; fully set with diamonds.
Strap: alligator leather.
Also available: in yellow gold; in pink gold; on gold bracelet.

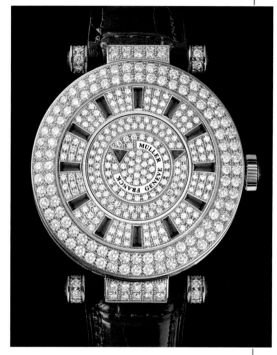

7045 GIGA TOURBILLON SQUELETTE ROND

Movement: manual-winding GIGA tourbillon caliber; skeletonized; 9-day power reserve; 4 barrels arranged in 2 pairs.
Functions: hours, minutes; tourbillon; power reserve indication at 12.
Case: pink gold; Ø 45mm; sapphire crystal; water resistant.
Dial: openworked.
Strap: alligator leather.
Note: the largest tourbillon in the world, with a 20mm carriage; a Franck Muller exclusive.
Also available: in white gold; in titanium; in steel.

7039 SQUELETTE RESERVE DE MARCHE 7 JOURS ROND – SERTIE BAGUETTES

Movement: manual-winding; skeletonized; 7-day power reserve.
Functions: hours, minutes; small seconds at 6.
Case: pink gold; Ø 39mm; set with baguette-cut diamonds; sapphire crystal; transparent caseback; water resistant to 30m.
Dial: openworked.
Strap: alligator leather.
Also available: in white gold; in platinum.

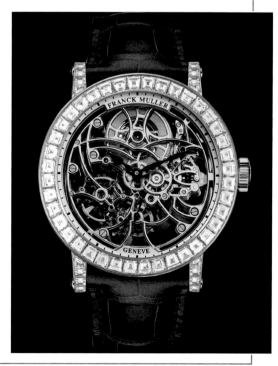

FRANCK MULLER

V32 LADY VANGUARD COLOR DREAM SERTIE

Movement: quartz.
Functions: hours, minutes.
Case: white gold; set with diamonds; sapphire crystal; water resistant to 30m.
Dial: white; sunray guilloché; "Color Dream" numerals; available in several colors.
Strap: alligator leather; natural rubber lining.
Also available: in pink gold; in steel.

V32 LADY VANGUARD SERTIE FULL

Movement: quartz.
Functions: hours, minutes.
Case: pink gold; set with diamonds; sapphire crystal; water resistant to 30m.
Dial: fully set with diamonds; numerals outlined in pink; available in several colors.
Strap: alligator leather; natural rubber lining.
Also available: in steel; in white gold.

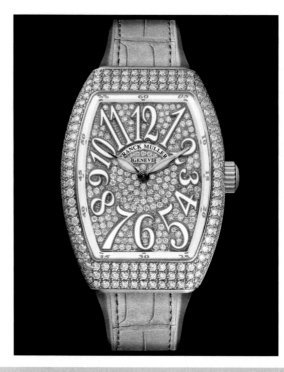

V45 SC DT VANGUARD

Movement: automatic-winding.
Functions: hours, minutes, seconds; date at 6.
Case: pink gold; brushed; sapphire crystal; water resistant to 30m.
Dial: black; brushed; numerals outlined in pink-gold relief; available in several colors.
Strap: alligator leather; natural rubber lining.
Also available: in steel; in white gold; in titanium.

V45 CC DT VANGUARD

Movement: automatic-winding chronograph.
Functions: hours, minutes; 2-counter chronograph; date at 6.
Case: titanium and 5N pink gold; "Vanguard"; interior pink-gold bezel; sapphire crystal; water resistant to 30m.
Dial: brushed titanium; white numerals outlined in pink-gold relief;
Strap: calfskin; natural rubber lining.
Also available: in titanium.

FRANCK MULLER

SQUELETTE RESERVE DE MARCHE 7 JOURS VANGUARD YACHTING

Movement: manual-winding; skeletonized; 7-day power reserve.
Functions: hours, minutes; small seconds in compass at 6.
Case: pink gold; sapphire crystal; transparent caseback; water resistant to 30m.
Dial: openworked; blue anchor yachting decoration at 12.
Strap: alligator leather; natural rubber lining.
Also available: in white gold; in steel.

GRAVITY TOURBILLON VANGUARD

Movement: manual-winding tourbillon carrousel caliber; skeletonized; 120-hour power reserve; Franck Muller exclusive.
Functions: hours, minutes; tourbillon carrousel.
Case: titanium; sapphire crystal; water resistant to 30m.
Dial: openworked.
Strap: alligator leather; natural rubber lining.
Also available: in white gold; in pink gold.

V50 RMT SQUELETTE

Movement: manual-winding tourbillon minute repeater; skeletonized; Franck Muller exclusive.
Functions: hours, minutes; tourbillon; minute repeater sounding the hours, quarter-hours and minutes on demand.
Case: pink gold; sapphire crystal; transparent caseback; water resistant to 30m.
Dial: openworked.
Strap: alligator leather; natural rubber lining.
Also available: in white gold.

V45 T SQT SAPPHIRE

Movement: manual-winding tourbillon; skeletonized.
Functions: hours, minutes; tourbillon.
Case: 100% transparent sapphire; sapphire crystal; sapphire crystal caseback; water resistant to 30m; Franck Muller exclusive.
Dial: openworked.
Strap: alligator leather; natural rubber lining.

GP
GIRARD-PERREGAUX
HAUTE HORLOGERIE SUISSE DEPUIS 1791

POETRY OF LAUREATO

First introduced in the 1970s, **THE LAUREATO COLLECTION CONTINUES TO BE A POWERFUL DRIVING FORCE FOR GIRARD-PERREGAUX TO THIS DAY**. Its octagonal bezel, integrated case, clous de Paris dial, and visually impactful bracelet make the line instantly recognizable, while its versatility lends itself to a range of complications and techniques.

Available in two sizes (38mm or 42mm) and two case materials (steel or rose gold), the Laureato Chronograph brings the collection's unique appeal to life with a bold, high-contrast aesthetic. Proud members of the Laureato collection, these chronographs are housed in the integrated case, seamlessly passing to the supple metal bracelet. An iconic octagonal bezel frames the strikingly attractive dial. Three subdials with snailed finishing punctuate the clous de Paris dial, underscoring the subtle way in which haute horology techniques can delineate the watch's functions, and a date aperture at 4:30 showcases an extra touch of utility. The smooth operation of its chronograph pushbuttons is characteristic of the attention to detail that Girard-Perregaux lavishes upon its timepieces, as is their octagonal shape, a visual echo of the distinctive bezel. Within beats the GP03300-0134/0136/0137, an automatic movement that beats at 28,800 vph and provides a 46-hour power reserve.

◀ **LAUREATO CHRONOGRAPH**
From the contrasting finishes on the dial to the alternating ones on the bracelet, every millimeter of the Laureato Chronograph is carefully considered for maximum visual intrigue.

GIRARD-PERREGAUX

The Laureato Flying Tourbillon Skeleton carves away everything extraneous to provide the wearer with a vertigo-inducing dive into the depths of an extraordinary piece.

Combining the technical prowess of the original Laureato (equipped with an ultra-precise quartz movement on its initial launch in 1975) with the more traditional demands of conventional complications—and the aesthetic showmanship of skeletonization—the Laureato Flying Tourbillon Skeleton combines several ambitious elements to encapsulate Girard-Perregaux's quest for perfection. Housed in a 42mm case in 18-karat white or rose gold, the timepiece carves away everything extraneous to provide the wearer with a vertigo-inducing dive into the depths of an extraordinary piece, delicate in appearance but built on solid technical foundations. The attenuated dial is little more than a chapter ring, revealing the flying tourbillon and its environs (the movement comprises 262 components) from the front even as the sapphire crystal caseback bares all on the other side. The automatic movement inside, the GP09520-0001, beats at 21,600 vph, providing a power reserve of approximately 50 hours.

The Laureato Perpetual Calendar takes a modern, somewhat unusual approach to an age-old complication. Rather than the traditional arrangement of evenly sized subdials and hands, this model arranges its indications in an asymmetrical style that hints at a broader design philosophy. The date, leap-year cycle, and days of the week each enjoy a uniquely sized area—with a snailed finish for the date and day against a clous de Paris backdrop—while the months appear through a large window, with a red arrow at 5 o'clock indicating the present. As a perpetual calendar, of course, the timepiece requires no correction of the date until March 1, 2100. In contrast to most other perpetual calendars, the Laureato Perpetual Calendar sets the date, month, and position in the leap-year cycle using the crown (adjustable in either direction), with a separate pushbutton for the day cycle. The automatic GP01800-0033 movement powers the ensemble at 28,800 vph, with a power reserve of 54 hours.

▲ **LAUREATO FLYING TOURBILLON SKELETON**
Carving away all but the essential, the Laureato Flying Tourbillon Skeleton puts its stunning flying tourbillon movement on display, shading its fixed elements with black PVD treatment.

▶ **LAUREATO PERPETUAL CALENDAR**
The sturdy stainless steel case of the Laureato Perpetaul Calendar boasts a diameter of 42mm, thickness of 11.84mm, and water resistance to 100m.

GIRARD-PERREGAUX

TOURBILLON WITH THREE GOLD BRIDGES — REF. 99280-52-000-BA6E

Movement: automatic-winding GP09400-0007 caliber; 60-hour power reserve; 263 components; 27 jewels; 21,600 vph.
Functions: hours, minutes; tourbillon; small seconds on the tourbillon at 6.
Case: pink gold; Ø 45mm, thickness: 13.78mm; sapphire crystal caseback; water resistant to 3atm.
Strap: black alligator leather; pink-gold triple-folding buckle.
Suggested price: $171,000

NEO-TOURBILLON WITH THREE BRIDGES SKELETON — REF. 99295-21-000-BA6A

Movement: automatic-winding GP09400-0011 caliber; 60-hour power reserve; 260 components; 27 jewels; 21,600 vph.
Functions: hours, minutes; tourbillon; small seconds on tourbillon at 6.
Case: titanium; Ø 45mm, thickness: 15.60mm; sapphire crystal caseback; water resistant to 3atm.
Strap: black alligator leather; titanium triple-folding buckle.
Suggested price: $145,000

MINUTE REPEATER TOURBILLON WITH BRIDGES — REF. 99820-21-001-BA6A

Movement: manual-winding GP09500-0003 caliber; 60-hour power reserve; 406 components; 37 jewels; 21,600 vph.
Functions: hours, minutes; tourbillon; minute repeater.
Case: titanium; Ø 45mm, thickness: 15.91mm; sapphire crystal caseback; water resistant to 3atm.
Strap: black alligator leather; titanium triple-folding buckle.
Suggested price: $374,000

MINUTE REPEATER TRI-AXIAL TOURBILLON — REF. 99830-21-000-BA6A

Movement: manual-winding GP09560-0001 caliber; 64-hour power reserve; 518 components; 54 jewels; 21,600 vph.
Functions: hours, minutes; tri-axial tourbillon; minute repeater.
Case: titanium; Ø 48mm, thickness: 21.24mm; sapphire crystal caseback; water resistant to 3atm.
Dial: sapphire rings.
Strap: black alligator leather; titanium triple-folding buckle.
Suggested price: $441,000

GIRARD-PERREGAUX

CLASSIC BRIDGES 40 MM — REF. 86005-52-001-BB6A

Movement: automatic-winding GP08600-0001 caliber; 48-hour power reserve; 207 components; 29 jewels; 21,600 vph.
Functions: hours, minutes.
Case: pink gold; Ø 40mm, thickness: 11.70mm; sapphire crystal caseback; water resistant to 3atm.
Dial: hours ring with suspended indexes.
Strap: black alligator leather; pink-gold triple-folding buckle.
Suggested price: $34,100

NEO BRIDGES — REF. 84000-21-001-BB6A

Movement: automatic-winding GP08400-0001 caliber; 54-hour power reserve; 208 components; 29 jewels; 21,600 vph.
Functions: hours, minutes.
Case: titanium; Ø 45mm, thickness: 12.18mm; sapphire crystal caseback; water resistant to 3atm.
Dial: hours ring with suspended indexes.
Strap: black alligator leather; titanium triple-folding buckle.
Suggested price: $25,200

1966 36 MM — REF. 49523-11-171-CB6A

Movement: automatic-winding GP03300-0123 caliber; 46-hour power reserve; 192 components; 27 jewels; 28,800 vph.
Functions: hours, minutes, seconds.
Case: steel; Ø 36mm, thickness: 9.01mm; sapphire crystal caseback; water resistant to 3atm.
Dial: flinqué silver.
Strap: black alligator leather; steel pin buckle.
Suggested price: $8,100

1966 40 MM — REF. 49555-11-131-BB60

Movement: automatic-winding GP03300-0030 caliber; 46-hour power reserve; 218 components; 27 jewels; 28,800 vph.
Functions: hours, minutes, seconds; date at 3.
Case: steel; Ø 40mm, thickness: 8.9mm; sapphire crystal caseback; water resistant to 3atm.
Dial: silver opaline.
Strap: black alligator leather; steel pin buckle.
Suggested price: $7,900

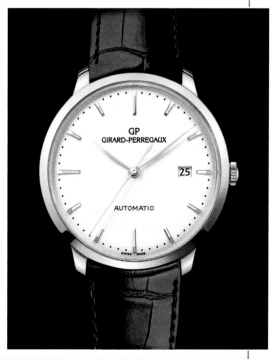

GIRARD-PERREGAUX

LAUREATO FLYING TOURBILLON SKELETON   REF. 99110-53-001-53A

Movement: automatic-winding GP09520-0001 caliber; 50-hour power reserve; 262 components; 28 jewels; 21,600 vph.
Functions: hours, minutes; flying tourbillon at 11.
Case: white gold; Ø 42mm, thickness: 10.76mm; sapphire crystal caseback; water resistant to 3atm.
Dial: hours ring with suspended indexes.
Bracelet: white gold; white-gold triple-folding buckle.
Suggested price: $136,000

LAUREATO SKELETON   REF. 81015-32-001-32A

Movement: automatic-winding GP01800-0006 caliber; 54-hour power reserve; 173 components; 25 jewels; 28,800 vph.
Functions: hours, minutes, small seconds.
Case: ceramic; Ø 42mm, thickness: 10.93mm; sapphire crystal caseback; water resistant to 10atm.
Dial: hour ring with suspended indexes.
Bracelet: ceramic; ceramic triple-folding buckle.
Suggested price: $38,400

LAUREATO CHRONOGRAPH 42 MM   REF. 81020-11-431-11A

Movement: automatic-winding GP03300-0137 caliber; 46-hour power reserve; 419 components; 63 jewels; 28,800 vph.
Functions: hours, minutes; small seconds; chronograph: 60-minute counter at 3, 12-hour counter at 6, date at 4:30.
Case: 904L steel; Ø 42mm, thickness: 12.01mm, water resistant to 10atm.
Dial: blue; with "Clou de Paris" pattern.
Bracelet: 904L steel; 904L steel triple-folding buckle.
Suggested price: $15,000

LAUREATO CHRONOGRAPH 42 MM   REF. 81020-11-131-11A

Movement: automatic-winding GP03300-0137 caliber; 46-hour power reserve; 419 components; 63 jewels; 28,800 vph.
Functions: hours, minutes; small seconds; chronograph: 60-minute counter at 3, 12-hour counter at 6; date at 4:30.
Case: 904L steel; Ø 42mm, thickness: 12.01mm; water resistant to 10atm.
Dial: silver with "Clou de Paris" pattern.
Bracelet: 904L steel; 904L steel triple-folding buckle.
Suggested price: $15,000

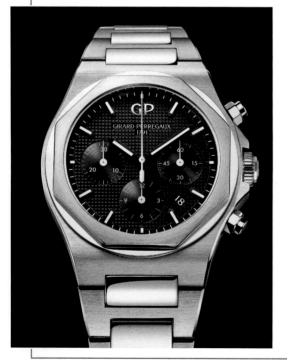

GIRARD-PERREGAUX

LAUREATO 42 MM   REF. 81010-11-431-11A

Movement: automatic-winding GP01800-0013 caliber; 54-hour power reserve; 191 components; 28 jewels; 28,800 vph.
Functions: hours, minutes, seconds; date at 3.
Case: steel; Ø 42mm, thickness: 10.88mm; sapphire crystal caseback; water resistant to 10 atm.
Dial: blue; with "Clou de Paris" pattern.
Bracelet: steel; steel triple-folding buckle.
Suggested price: $11,600

LAUREATO 42 MM   REF. 81010-32-631-32A

Movement: automatic-winding GP01800-0025 caliber; 54-hour power reserve; 191 components; 28 jewels; 28,800 vph.
Functions: hours, minutes, seconds; date at 3.
Case: ceramic; Ø 42mm, thickness: 10.93mm; sapphire crystal; sapphire crystal caseback; water resistant to 10 atm.
Dial: black; with "Clou de Paris" pattern.
Bracelet: ceramic; ceramic triple-folding buckle.
Suggested price: $16,500

LAUREATO 38 MM   REF. 81005D82A732-32A

Movement: automatic-winding GP03300-0030 caliber; 46-hour power reserve; 218 components; 27 jewels; 28,800 vph.
Functions: hours, minutes, central seconds; date at 3.
Case: ceramic; Ø 38mm, thickness: 10.20mm; steel bezel set with 56 brilliant-cut diamonds; sapphire crystal caseback; water resistant to 10atm.
Dial: white; with "Clou de Paris" pattern.
Bracelet: ceramic; ceramic triple-folding buckle.
Suggested price: $18,100

LAUREATO 34 MM   REF. 80189D56A331-56A

Movement: quartz GP013100-0002; 90 components; 7 jewels; 32,768 Hz.
Functions: hours, minutes; date at 3.
Case: steel and pink gold set with 56 brilliant-cut diamonds; Ø 34mm, thickness: 7.75mm; water resistant to 3 atm.
Dial: golden; with "Clou de Paris" pattern.
Bracelet: steel and pink gold.
Suggested price: $15,900

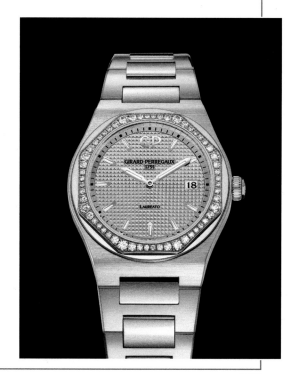

GUY ELLIA

A RING OF ENDLESS LIGHT

Curving to conform precisely to the feminine wrist, Guy Ellia's Circle "La Petite" suggests **THE BEAUTY OF INDIVIDUALITY**.

The case's unusual convex shape hints at the watchmaker's history as something of an iconoclast, while at 45mm in diameter, its "La Petite" nickname seems inapt, until one compares its dimensions to the brand's other larger-than-life models.

This stunning timepiece finds striking beauty in geometric simplicity. A variety of tones and gem settings complements its elegant shape to match the distinctive personality of its stylish wearer. Whether dressed in white, yellow or rose gold, the Circle "La Petite" highlights its captivating curvature with 105 brilliant-cut diamonds on the bezel that affirm the luxurious refinement of the design. Its thickness, of only 6mm, contrasts with the otherwise imposing dimensions to maintain a seamless tone of unassuming grace. On the dial, 12 elongated indexes, with or without 192 additional brilliant-cut stones, mark the hours and appear to radiate from a small central circle where the hours and minutes are told via two dauphine hands of just 6 and 7.5mm. The Circle "La Petite" also expresses itself with a wide variety of dial colors. From brilliant black to opal, night blue, polished or matte "rose" and polished or matte yellow, the timepiece allows its owner to make an uncompromising aesthetic statement. The watch may thus range from a single-tone demonstration of nuanced subtleties to a high-contrast creation full of bold combinations and shimmering accents.

The Circle "La Petite" has yet another shimmering trick up its sleeve. A full-snow-set model replaces its colorful dial and converging rays with an incredible 1,044 brilliant-cut diamonds, totalling 4.88 carats. Gracing its riveting convex disposition with boundless light and intriguing texture, this snow-set variation reflects the bold individuality of its wearer with sumptuous opulence.

The Circle "La Petite" is finished with a 2.8mm diamond on the crown and is secured to the wrist by way of a curved bracelet perfectly integrated into the structure of the timepiece.

▶ **CIRCLE "LA PETITE"**
Available in three case choices and with a variety of dial personalities and gem settings, this convex ladies' wristwatch showcases its unique personality at every turn.

The slender 6mm thickness of the Circle "La Petite" contrasts with the piece's otherwise imposing dimensions to maintain a seamless tone of unassuming grace.

GUY ELLIA

ELYPSE

Movement: quartz; Blancpain caliber PGE 820; Ø 18.8mm, thickness: 1.95mm; rhodiumed and "Côtes de Genève"-decorated cage.
Functions: hours, minutes.
Case: 18K white gold (53.38g); 52x35mm, thickness: 5.6mm; bezel set with 80 diamonds; crown set with one diamond (Ø 2.5mm); sapphire glass with thermal counter-shock marking; mirror-polished caseback set with one diamond on the "I" of ELLIA (Ø 0.95mm) and deep mechanical engraving; water resistant to 3atm.
Dial: shiny gold black painted; Roman numerals; gold dauphine-shaped hands.

Strap: black alligator leather; 18K white-gold pin buckle set with nine diamonds (Ø 1.5mm); pin set with one diamond (Ø 0.9mm) (0.12 carat).
Also available: case: pink and yellow gold; dial: Roman numeral versions: matte black, matte chocolate, brilliant gold, matte khaki, brilliant Burgundy, matte Burgundy, brilliant blue, matte orange, matte lilac, pink salmon, matte beige; dial: mirror or polished gold (white, pink and yellow), brilliant black, black matte, pearly white, light brown matte, navy pearly, light blue matte, opal.

ELYPSE

Movement: quartz; Blancpain caliber PGE 820; Ø 18.8mm, thickness: 1.95mm; rhodiumed and "Côtes de Genève"-decorated cage.
Functions: hours, minutes.
Case: white gold (53.38g); 52x35mm, thickness: 5.6mm; bezel set with 80 diamonds; crown set with one diamond (Ø 2.5mm); sapphire glass with thermal counter-shock marking; mirror-polished caseback set with one diamond on the "I" of ELLIA (Ø 0.9mm) and deep mechanical engraving; water resistant to 3atm.
Dial: microblasted white gold; blue painted Roman numerals; gold dauphine-shaped hands.

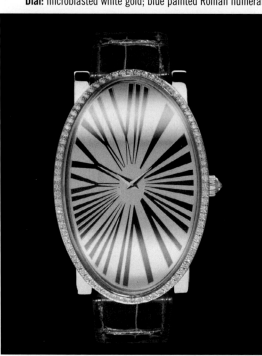

Strap: blue alligator leather; 18K white-gold buckle set with nine diamonds (Ø 1.5mm); pin set with one diamond (Ø 0.9mm) (0.12 carat).
Also available: case: pink and yellow gold; dial: Roman numeral versions: matte black, matte chocolate, brilliant gold, matte khaki, brilliant Burgundy, matte Burgundy, matte orange, matte lilac, pink salmon, matte beige; dial: mirror or polished gold (white, pink and yellow); brilliant black, black matte, light brown matte, navy pearly, light blue matte, opal.

QUEEN

Movement: quartz; Blancpain caliber PGE 820; Ø 18.8mm, thickness: 1.95mm; rhodiumed and "Côtes de Genève"-decorated cage.
Functions: hours, minutes.
Case: white gold (63.13g); 52x38.5mm, thickness: 7.2mm; bezel set with 96 diamonds (0.89 carat); crown set with one diamond (0.08 carat); sapphire glass with thermal counter-shock marking; mirror-polished caseback set with one diamond on the "I" of ELLIA (Ø 0.95mm) and deep mechanical engraving; water resistant to 3atm.
Dial: navy mother-of-pearl; white-gold Roman numerals.

Strap: black alligator leather; solid 18K white-gold pin buckle (4.32g) set with 81 diamonds (0.37 carat).
Also available: case: pink gold, yellow gold; mother-of-pearl dial with Roman numerals or mirror-polished dial with indexes set with 116 brillant-cut diamonds (0.35 carat)

QUEEN

Movement: quartz; Blancpain caliber PGE 820; Ø 18.8mm, thickness: 1.95mm; rhodiumed and "Côtes de Genève"-decorated cage.
Functions: hours, minutes.
Case: 18K pink gold (59.34g); 52x38.5mm, thickness: 7.2mm; fully set with 409 diamonds (3.05 carats); crown set with one diamond (0.08 carat); sapphire glass with thermal counter-shock marking; mirror-polished caseback set with one diamond on the "I" of ELLIA (Ø 0.95mm) and deep mechanical engraving; water resistant to 3atm.
Dial: pink gold; fully set with 450 diamonds (2.19 carats).

Strap: brown alligator leather; solid 18K pink-gold buckle and pin (3.84g) set with 81 diamonds (0.37 carat).
Also available: case: pink gold, yellow gold; mother-of-pearl dial with Roman numals or mirror-polished dial with indexes set with 116 brillant-cut diamonds (0.35 carat)

JUMBO CHRONO

Movement: automatic-winding Blancpain caliber PGE 1185; Ø 25.6mm, thickness: 5.5mm; 45-hour power reserve; chronograph with column wheel; Côtes de Genève finished bridges with rhodium plating; "GUY ELLIA" engraved on rotor with rhodium plating.
Functions: hours, minutes at 12; small seconds at 6; date at 2; column-wheel chronograph: 12-hour counter at 8, 30-minute counter at 4, central sweep seconds hand.
Case: 18K pink gold (85.75g); Ø 50mm, thickness: 11.5mm; sapphire crystal with thermal antireflective coating; water resistant to 3atm.
Strap: brown alligator leather; 18K pink-gold folding buckle (11.90g).
Also available: black gold; set black-gold bezel; full-set black gold; white gold; set white-gold bezel; full-set white gold; set pink-gold bezel; full-set pink gold.

JUMBO CHRONO

Movement: automatic-winding Blancpain caliber PGE 1185; Ø 25.6mm, thickness: 5.5mm; 45-hour power reserve; chronograph with column wheel; Côtes de Genève finished bridges with rhodium plating; "GUY ELLIA" engraved on rotor with rhodium plating.
Functions: hours, minutes at 12; small seconds at 6; date at 2; column-wheel chronograph: 12-hour counter at 8, 30-minute counter at 4, central sweep seconds hand.
Case: microblasted 18K white gold (90.14g); Ø 50mm, thickness: 11.5mm; bezel full set with 323 diamonds (7.93 carats); sapphire crystal with antireflective coating; water resistant to 3atm.
Strap: gray alligator leather; 18K white-gold folding buckle; set with 35 diamonds (0.35 carat).
Also available: black gold; set black-gold bezel; full-set black gold; white gold; set white-gold bezel; pink gold; set pink-gold bezel; full-set pink gold.

JUMBO HEURE UNIVERSELLE

Movement: automatic-winding Blancpain caliber PGE 1150; Ø 36.2mm, thickness: 6.24mm; 72-hour power reserve; 37 jewels; five-position adjustment; blue sapphire disc; Côtes de Genève-finished bridges with rhodium plating; "GUY ELLIA" engraved on rotor with rhodium plating.
Functions: hours, minutes; 24-hour time zone indicator; large date; day/night indicator.
Case: microblasted 18K gray gold (82.9g); Ø 50mm, thickness: 11mm; sapphire crystal with antireflective coating; openwork caseback with sapphire crystal; water resistant to 3atm.
Strap: gray alligator leather; 18K gray-gold folding buckle (16g).
Also available: white gold; black gold.

JUMBO HEURE UNIVERSELLE

Movement: automatic-winding Blancpain caliber PGE 1150; Ø 36.2mm, thickness: 6.24mm; 72-hour power reserve; 37 jewels; five-position adjustment; blue sapphire disc; Côtes de Genève-finished bridges with rhodium plating; "GUY ELLIA" engraved on rotor with rhodium plating.
Functions: hours, minutes; 24-hour time zone indicator; large date; day/night indicator.
Case: microblasted 18K black gold (82.9g); Ø 50mm, thickness: 11mm; sapphire crystal with antireflective coating; openwork caseback with sapphire glass; water resistant to 3atm.
Strap: black alligator leather; 18K black-gold folding buckle (16g).
Also available: white gold; pink gold.

GUY ELLIA

TOURBILLON ZEPHYR

Movement: manual-winding Christophe Claret caliber GES 97; 37x37mm, thickness: 6.21mm; 110-hour power reserve; 233 components; 17 jewels; 21,600 vph; winding ring set with 36 baguette-cut diamonds (1.04 carats) or engine turning; entirely hand-chamfered cage; bottom plate and bridges in blue sapphire; five-position adjustment.
Functions: hours, minutes; tourbillon.
Case: white sapphire case with platinum sides (54.9g); 54x45.3mm, thickness: 15.4mm; crown set with one diamond (Ø 1mm); water resistant to 3atm.
Strap: alligator leather; white-gold folding buckle (15.64g).

Note: limited edition of 12 number pieces.
Also available: pink gold; bottom plate and bridges sapphire: white, smokey.

REPETITION MINUTE ZEPHYR

Movement: manual-winding Christophe Claret GEC 88 caliber; 41.2x38.2mm, thickness: 9.41mm; 48-hour power reserve; 720 components; 72 jewels; 18,000 vph; gear wheels with different platings; five-position adjustment.
Functions: hours, minutes; power reserve indicator; minute repeater; five time zones with day/night indicators.
Case: sapphire crystal block and titanium; 53.6x43.7mm, thickness: 14.8mm; sapphire and titanium crown set with a diamond (Ø 2.2mm); water resistant to 3atm.
Strap: black rubber with titanium and white-gold folding buckle (17.27g).

Note: limited edition of 20 numbered pieces.
Also available: pink gold and white gold; alligator leather strap.

TOURBILLON MAGISTERE

Movement: manual-winding Christophe Claret caliber PGE 97; 37.4x29.9mm, thickness: 5.4mm; 110-hour power reserve; 20 jewels; 21,600 vph; flat balance-spring; mysterious winding; skeletonized barrel and ratchet wheel; entirely hand-chamfered cage; white-gold bottom plate and bridges.
Functions: hours, minutes; tourbillon.
Case: pink gold (73.09g); 43.5x36mm, thickness: 10.9mm; sapphire glass with anti-reflective coating; water resistant to 3atm.
Strap: brown alligator leather; 18K pink-gold folding buckle (14.61g).

Also available: white gold; titanium; platinum; bezel set with 52 baguette diamonds (2.15 carats); full set with 172 baguette diamonds (19.56 carats); full set with 535 brilliant-cut diamonds (8.25 carats).

TOURBILLON MAGISTERE II

Movement: manual-winding Christophe Claret caliber MGE 97; 38.4x30.9mm, thickness: 5.71mm; 90-hour power reserve; 266 components; 33 jewels; 21,600 vph; flat balance-spring; mysterious winding; skeletonized ratchet and wheels with curved arms and wolf-teeth; entirely hand-chamfered cage; 18K white-gold tourbillon and barrel bridges.
Functions: hours, minutes; tourbillon.
Case: white gold (125g); 44x36.7mm, thickness: 15mm; antireflective sapphire glass; transparent caseback; water resistant to 3atm.
Strap: black alligator leather; solid 18K white-gold folding buckle (18.77g).
Note: limited edition of 12 numbered pieces.
Also available: red gold.

TIME SPACE

Movement: manual-winding Blancpain caliber PGE 15; Ø 35.64mm, thickness: 1.9mm; 43-hour power reserve; 20 jewels; 21,600 vph; five position adjustment; Côtes de Genève finished bridges with black PVD treatment; "GE" engraved on stippled plate with black PVD treatment.
Functions: hours, minutes.
Case: 18K full-set black gold (33.73g); Ø 46.8mm, thickness: 4.9mm; bezel set with 234 diamonds (1.83 carats); bottom plate set with 366 diamonds (1.03 carats); sapphire crystal with antireflective coating; water resistant to 3atm.
Strap: black alligator leather; 18K black-gold pin buckle (4.82g) set with 31 diamonds (0.41 carat).
Also available: white gold; set-bezel white gold; full-set white gold; pink gold; set-bezel pink gold; full-set pink gold; black gold; set-bezel black gold.

TIME SPACE

Movement: manual-winding Blancpain caliber PGE 15; Ø 35.64mm, thickness: 1.9mm; 43-hour power reserve; 20 jewels; 21,600 vph; five position adjustment; Côtes de Genève finished bridges with black PVD treatment; "GE" engraved on stippled plate with black PVD treatment.
Functions: hours, minutes.
Case: 18K pink gold (31.42g); Ø 46.8mm, thickness: 4.9mm; sapphire glass with antireflective coating; water resistant to 3atm.
Strap: brown alligator leather; gold pin buckle (4.28g).
Also available: white gold; set-bezel white gold; full-set white gold; set-bezel pink gold; full-set pink gold; black gold; set-bezel black gold; full-set black gold.

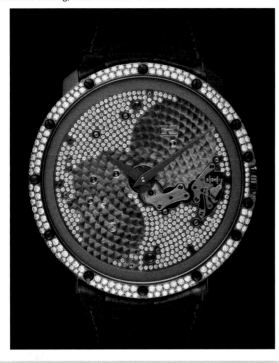

TIME SPACE QUANTIEME PERPETUEL

Movement: manual-winding Blancpain PGE 5615 D caliber; Ø 35.64mm, thickness: 4.7mm; 43-hour power reserve; 20 jewels; 21,600 vph; five-position adjustment; Côtes de Genève-finished bridges with black PVD treatment; GUY ELLIA logo engraved on stippled plate with black PVD treatment; watch box with an integrated specific automatic winder.
Functions: hours, minutes; perpetual calendar: day, date, month, leap-year cycle; moonphase.
Case: black gold (32.61g); Ø 46.8mm, thickness: 7.75mm; crown set with one brilliant-cut diamond (Ø 1.3mm); sapphire middle ring; sapphire crystal with antireflective coating; water resistant to 3atm.
Strap: black alligator leather; 18K black-gold pin buckle (4.82g).
Also available: black-gold bezel set, full-set black gold; white gold, white-gold bezel set, full-set white gold; pink gold, pink-gold bezel set, full-set pink gold.

TIME SPACE QUANTIEME PERPETUEL

Movement: manual-winding Blancpain caliber PGE 15; Ø 35.64mm, thickness: 4.7mm; 43-hour power reserve; 20 jewels; 21,600 vph; five-position adjustment; Côtes de Genève finished bridges with black PVD treatment; "GE" engraved on stippled plate with black PVD treatment; watch box with an integrated specific automatic winder.
Functions: hours, minutes; perpetual calendar: day, date, month, leap-year cycle; moonphase.
Case: white gold (32.61g); Ø 46.8mm, thickness: 7.75mm; crown set with one diamond (Ø 1.3mm); sapphire middle ring; sapphire crystal with antireflective coating; water resistant to 3atm.
Strap: black alligator leather; 18K white-gold pin buckle (4.82g).
Also available: set-bezel white gold; full-set white gold; pink gold; set-bezel pink gold; full-set pink gold; black gold; set-bezel black gold; full-set black gold.

GUY ELLIA

DOUZE

Movement: quartz; Blancpain caliber PGE 820; Ø 18.8mm, thickness: 1.95mm; rhodiumed and "Côtes de Genève"-decorated cage.
Functions: hours, minutes.
Case: 18K pink gold (74.44g); 52.5x35.5mm, thickness: 5.8mm; bezel set with 82 diamonds (Ø 2mm); crown set with one diamond (Ø 2.3mm); sapphire glass with thermal counter-shock marking; mirror-polished caseback set with one diamond on the "I" of ELLIA (Ø 0.95mm) and deep mechanical engraving; water resistant to 3atm.
Strap: navy alligator leather; 18K pink-gold pin buckle set with 13 diamonds (Ø 2mm); pin set with one diamond (Ø 0.9mm).
Also available: case and pin buckle: yellow gold, white gold; dial: opal, gray, black, chocolate, burgundy; markers: mirror-polished gold (yellow, white).

DOUZE

Movement: quartz; Blancpain caliber PGE 820; Ø 18.8mm, thickness: 1.95mm; rhodiumed and "Côtes de Genève"-decorated cage.
Functions: hours, minutes.
Case: 18K pink gold (74.44g); 52.5x35.5mm, thickness: 5.8mm; bezel set with 82 diamonds (Ø 2.5mm); crown set with one diamond (Ø 2.3mm); sapphire glass with thermal counter-shock marking; mirror-polished caseback set with one diamond on the "I" of ELLIA (Ø 0.95mm) and deep mechanical engraving; water resistant to 3atm.
Strap: light brown alligator leather; 18K pink-gold pin buckle set with 13 diamonds (Ø 2mm); pin set with one diamond (Ø 0.9mm).
Also available: case and pin buckle: yellow gold, white gold; dial: navy, gray, black, chocolate, burgundy; markers: mirror-polished gold (yellow, white).

CONVEX

Movement: quartz; Blancpain caliber PGE 820; Ø 18.8mm, thickness: 1.95mm; rhodiumed and "Côtes de Genève"-decorated cage.
Functions: hours, minutes.
Case: 18K white gold (59.17g); 41x41mm, thickness: 7.1mm; bezel set with 100 diamonds (Ø 1.5mm); crown set with one diamond (Ø 2.3mm) (1.25 carats); sapphire glass with thermal counter-shock marking; mirror-polished caseback set with one diamond on the "I" of ELLIA (Ø 0.95mm) and deep mechanical engraving; water resistant to 3atm.
Dial: 18K white gold (25.35g); fully set with 558 diamonds (2.28 carats); four gold Arabic numerals; 18K white-gold dauphine-shaped hands.
Strap: black alligator leather; 18K white-gold pin buckle set with 40 diamonds (Ø 1.5mm); pin set with one diamond (Ø 0.9mm) (0.49 carat).
Also available: case: yellow gold, pink gold; dial: solid 18K gold (white, yellow, pink), mother-of-pearl (navy, jeans, pink beige); markers: diamond set with 204 diamonds (0.79 carat), gold (white, yellow, pink), outlined.

CONVEX

Movement: quartz; Blancpain caliber PGE 820; Ø 18.8mm, thickness: 1.95mm; rhodiumed and "Côtes de Genève"-decorated cage.
Functions: hours, minutes.
Case: 18K pink gold (55.4g); 41x41mm, thickness: 7.1mm; bezel set with 100 diamonds (Ø 1.5mm); crown set with one diamond (Ø 2.3mm); sapphire glass with thermal counter-shock marking; mirror-polished caseback set with one diamond on the "I" of ELLIA (Ø 0.95mm) and deep mechanical engraving; water resistant to 3atm.
Dial: mother-of-pearl with four Arabic numerals; 18K gold dauphine-shaped hands.
Strap: dark blue alligator leather; 18K pink-gold pin buckle set with 40 diamonds (Ø 1.5mm); pin set with one diamond (Ø 0.9mm) (0.49 carat).
Also available: case: yellow gold, white gold; dial: fully set 18K gold (white, yellow, pink), mother-of-pearl: navy, jeans, pink beige; markers: diamond set with 204 diamonds (0.79 carat), gold (white, yellow, pink), outlined.

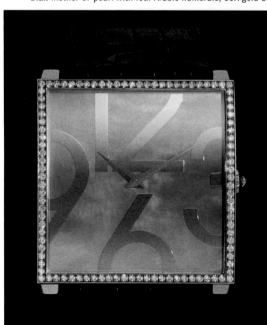

CIRCLE "LA PETITE"

Movement: quartz; Blancpain caliber PGE 820; Ø 18.8mm, thickness: 1.95mm.
Functions: hours, minutes.
Case: polished white gold (41g); Ø 45mm, thickness: 6mm; bezel set with 105 diamonds (0.71 carat); crown set with one diamond (Ø 2.8mm) (0.08 carat); sapphire crystal with thermal counter-shock marking; caseback set with one diamond on the "I" of ELLIA (Ø 0.95mm) (0.003 carat); water resistant to 3atm.
Dial: mirror white; hour markers set with 192 brilliant-cut diamonds (0.58 carat); 18K white-gold dauphine-shaped hands.
Bracelet: solid 18K white gold (115.1g).
Also available: case: pink gold, yellow gold; dial: shiny black, opal or navy, mirror or matte gold white, pink and yellow; markers: mirror-polished gold, diamond set; strap: alligator leather with 18K solid gold pin buckle set with one diamond (Ø 0.9mm).

CIRCLE "LA PETITE"

Movement: quartz; Blancpain caliber PGE 820; Ø 18.8mm, thickness: 1.95mm.
Functions: hours, minutes.
Case: pink gold (37g); Ø 45mm, thickness: 6mm; bezel set with 105 diamonds (0.71 carat); crown set with one diamond (Ø 2.8mm); sapphire crystal with thermal counter-shock marking; caseback set with one diamond on the "I" of ELLIA (Ø 0.95mm) (0.003 carat); water resistant to 3atm.
Dial: black; pink-gold mirror-polished hour markers; 18K pink-gold dauphine-shaped hands.
Bracelet: pink gold (104.2g).
Also available: case: pink gold, yellow gold, white gold; dial: opal, navy, matte gold, full set; markers: mirror-polished gold, diamond set; strap: alligator leather with 18K solid gold pin buckle set with one diamond (Ø 0.9mm).

CIRCLE

Movement: quartz; Blancpain caliber PGE 820; Ø 18.8mm, thickness: 1.95mm.
Functions: hours, minutes.
Case: pink gold (80.13g); Ø 52mm, thickness: 7mm; bezel set with 124 diamonds (Ø 1.15mm); crown set with one diamond (Ø 2.8mm); sapphire crystal with thermal counter-shock marking; mirror-polished caseback set with one diamond on the "I" of ELLIA (Ø 0.95mm); water resistant to 3atm.
Dial: opal; pink-gold hour markers; 18K pink-gold dauphine-shaped hands.
Strap: alligator leather; solid 18K pink-gold pin buckle set with 86 diamonds (0.384 carat); pin set with one diamond (Ø 0.9mm).
Also available: case: black gold, yellow gold, white gold; dial: opal, mirror or matte gold (white, pink, yellow), full set; markers: diamond set, full set.

CIRCLE

Movement: quartz; Blancpain caliber PGE 820; Ø 18.8mm, thickness: 1.95mm.
Functions: hours, minutes.
Case: black gold (83.86g); Ø 52mm, thickness: 7mm; bezel set with 124 diamonds (0.743 carat); crown set with one diamond (Ø 2.8mm); sapphire crystal with thermal counter-shock marking; caseback set with one diamond on the "I" of ELLIA (Ø 0.95mm); water resistant to 3atm.
Dial: matte black; hour markers set with 168 brilliant-cut diamonds (1.0025 carats); 18K white-gold dauphine-shaped hands.
Strap: alligator leather; solid 18K black-gold pin buckle set with 86 diamonds (0.384 carat); pin set with one diamond (Ø 0.9mm).
Also available: case: pink gold, yellow gold, white gold; dial: opal, mirror or matte gold (black, white, pink, yellow), full set; markers: gold, full set.

HUBLOT

ENDLESS ADVENTURES

As committed as ever to its mission statement, "The Art of Fusion," Hublot releases watches that **CLEVERLY MELD AND TRANSFORM REFERENCES**. Whether gentlemen's fashion, dangerous luxury, beachy natural beauty, cozy retreats, or the nature of starstuff is on the table, these models incorporate broadly ranging touchstones in a unified design philosophy.

Solidifying its bona fides with the fashion industry, Hublot once again partners with Italia Independent for an eye-opening take on houndstooth, on its Classic Fusion Chronograph Italia Independent Pied-De-Poule Blue Titanium. Woven into the Rubinacci suit fabric, the timeless motif brings a masculine edge to the proceedings, in two daring shades of blue. The soft material, which covers the strap and dial, runs into the satin-finished and polished titanium of the 45mm case, equipped with two chronograph pushers and Hublot's signature H-shaped titanium screws. Powered by the HUB1143 self-winding movement, the model shows off its chronograph function with a 30-minute counter at 9 o'clock and a central chronograph hand, complementing the small seconds display at 3 o'clock. The combination of expertly woven fabric with precise clockwork and polished metal provides a fresh perspective on elegant masculinity.

Also in collaboration with Italia Independent, Hublot has created the Classic Fusion Chronograph Italia Independent Pinstripe Ceramic, which brings the narrow gentlemen's fashion mainstay to the wrist. Boasting the reliable efficiency of the HUB1143 movement, which comes with a 42-hour power reserve, this chronograph is a boardroom bombshell, recreating the look of a business suit in Hublot's inimitable cutting-edge style. The 45mm case in satin-finished and polished black ceramic adds to the muted, all-business tone of the timepiece. A central chronograph seconds hand swings about the perfectly symmetrical dial, with two evenly matched counters for the small seconds (at 3 o'clock) and 30-minute counter (at 9 o'clock). The dial and strap are covered in authentic Rubinacci pinstripe, with black stitching and blue rubber lining on the strap.

Three hundred and thirty-six diamonds on the case and dial lend a reptilian shimmer to the extravagant Big Bang Tourbillon Croco High Jewelry.

As intriguing as understatement can be, sometimes it is to be avoided at all costs. Hublot's Big Bang Tourbillon Croco High Jewelry is coated in diamonds from lugs to dial. Two hundred and thirty-four diamonds (totaling 7.6 carats) on the 45mm case, as well as 102 diamonds (totaling 4.3 carats) on the dial, lend a reptilian shimmer to this extravagant model. Baguette-cut in varying shapes, the diamonds cover the watch in an irregular pattern, resembling the dangerous look of a crocodile's belly. The motif extends to the crocodile strap, whose clasp also bears 44 diamond baguettes. The HUB6016 movement beating inside powers an entrancing skeletonized tourbillon, drawing upon a generous power reserve of approximately 115 hours. An aperture on the dial at 9 o'clock keeps track of how much autonomy remains in this stunning creation.

▶ **BIG BANG TOURBILLON CROCO HIGH JEWELRY**
Set with 13.5 carats of diamonds that create a stunning reptilian effect, the Big Bang Tourbillon Croco High Jewelry uses jewelry expertise to highlight impeccable watchmaking.

◀ **CLASSIC FUSION CHRONOGRAPH ITALIA INDEPENDENT PIED-DE-POULE BLUE TITANIUM** *(facing page top)*
Combining classic houndstooth and Hublot's signature style, this Classic Fusion model is a must-have for the sharp dressed man.

◀ **CLASSIC FUSION CHRONOGRAPH ITALIA INDEPENDENT PINSTRIPE CERAMIC** *(facing page bottom)*
With a bold pinstripe motif, the Classic Fusion Chronograph Italia Independent Pinstripe Ceramic takes care of business with Hublot's signature philosophy.

HUBLOT

▲ **CLASSIC FUSION AEROFUSION CHRONOGRAPH CAPRI**
Imbued with the serenity and drama of Capri's unforgettable natural beauty, the Classic Fusion Aerofusion Chronograph Capri is a piece of summer that lasts all year round.

▼ **BIG BANG ALPS**
From the ice-blue dial, to the granite-like case and bezel, to the cozy removable cuff, Hublot's Big Bang Alps exudes après-ski chic.

Like a cool Mediterranean breeze, the Classic Fusion Aerofusion Chronograph Capri is a refreshing piece of summer. The 45mm polished, satin-finished ceramic case captures a bit of the turquoise splendor of the water surrounding that dreamlike isle, with an extra splash of color on the central chronograph seconds hand. Available as a limited and numbered edition of 30 pieces for Hublot customers on the island, the Classic Fusion Aerofusion Chronograph Capri bears a fitting engraving on the polished ceramic caseback: the Faraglioni rocks that jut out of the water, a dramatic seascape that no visitor will ever forget. Behind the openworked dial, the HUB1155 self-winding chronograph movement provides a power reserve of 42 hours. A white lined rubber strap completes the carefree Capri look.

Celebrating a completely different but equally compelling landscape, Hublot's Big Bang Alps sweeps us off to the skiing capital of the world. The frosted carbon fiber construction of the 45mm Big Bang case evokes the rime that creeps up the outside of a window of a Swiss chalet, entrancingly beautiful to the people inside enjoying a bit of après-ski, as well as the sturdy granite that supports the region's mountains. The matte ice-blue openworked dial features a 60-minute chronograph counter at 3 o'clock and a subtle surprise in the small seconds counter at 9 o'clock: an appliqué of the towering Matterhorn, Switzerland's most famous mountain and the ultimate inspiration for this watch. The removable shearling "Cuddly Cuff" adds an extra touch of Alpine chic to the ice-blue sheep leather strap.

HUBLOT

Hublot presents a masterful command of simplicity with its Classic Fusion King Gold Green. The satin-finished green sunray dial, complete with a small seconds display and 30-minute counter, matches the olive green alligator leather strap. Crafted in Hublot's proprietary 18-karat King Gold alloy, the case and bezel complement the green color motif with a striking, unique golden hue. The HUB1143 self-winding movement inside powers the time display, chronograph functions, and a small date display at 6 o'clock at 28,800 vph. This sleek piece reveals the clean, minimalist foundation from which the rest of Hublot's empire extends.

▶ **CLASSIC FUSION KING GOLD GREEN**
With clean lines and Hublot's own King Gold alloy, this Classic Fusion King Gold Green reveals the brand's emphasis on impeccable watchmaking fundamentals.

In an act of radical transparency, Hublot lays it all bare with the Big Bang Unico Sapphire Galaxy. The 45mm case is created from polished sapphire crystal, and the skeletonized dial matches its transparency for a full view of the column-wheel chronograph HUB1242 UNICO movement within. A burst of color around the dial uses the bright hues of baguette-cut amethysts, rose sapphires, orange sapphires, fuchsia sapphires, Madagascar sapphires, and blue sapphires to evoke the extraordinary colors seen in astronomical photographs of interstellar nebulae. Luminescent white appliqués on the hands and hour markers add to the model's otherworldly beauty. A transparent structured lined rubber strap completes the look.

◀ **BIG BANG UNICO SAPPHIRE GALAXY**
A total of 48 amethysts and colored sapphires encircle the dial of the Big Bang Unico Sapphire Galaxy, framing the openworked dial, which possesses a 60-minute counter and date display at 3 o'clock, small seconds at 9 o'clock, and a central chronograph hand.

HUBLOT

BIG BANG REFEREE 2018 FIFA WORLD CUP RUSSIA — REF. 400.NX.1100.RX

Movement: powered by INTEL® Wear OS by Google technology developed by LVMH; 1-day battery life.
Functions: hours, minutes; apps: incoming calls, messages, referee app, chrono app, fitness, weather, translator, and maps; game mode during matches: countdown, game time alerts, goals, half-time, player substitution, red & yellow cards, additional time, and end of game; compatible with Android 4.4 and above, iOS 9 and above, and iPhone 5 or more recent.
Case: polished and satin-finished titanium; Ø 35.4mm; water resistant to 5atm.
Dial: 400x400 pixels; 287 ppi Amoled.
Strap: black structured lined rubber; additional sponge sport strap with HUBLOT logo and 2018 World Cup logo.
Note: watch can be customized with dials and straps for your favorite team.
Suggested price: $5,200

BIG BANG UNICO RED MAGIC — REF. 411.CF.8513.RX

Movement: automatic-winding HUB1242 UNICO Manufacture caliber.
Functions: hours, minutes; small seconds at 9; date at 3; chronograph: central seconds hand, 60-minute counter at 3.
Case: polished red ceramic.
Dial: skeletonized matte red.
Strap: black and red structured lined rubber.
Suggested price: $26,200

BIG BANG UNICO GMT KING GOLD — REF. 471.OX.7128.RX

Movement: automatic-winding HUB1251 UNICO Manufacture caliber.
Functions: hours, minutes, seconds; GMT; day/night indicator.
Case: satin-finished and polished 18K King Gold.
Dial: skeletonized blue and gold-plated.
Strap: black and blue lined structured rubber.
Suggested price: $40,900

BIG BANG FERRARI CARBON RED CERAMIC — REF. 402.QF.0110.WR

Movement: automatic-winding HUB1241 UNICO Manufacture caliber.
Functions: hours, minutes; small seconds at 9; date at 3; chronograph: central seconds hand, 60-minute counter at 3.
Case: unidirectional carbon fiber; polished red ceramic bezel.
Dial: sapphire crystal.
Strap: black rubber and alcantara.
Suggested price: $29,400

HUBLOT

MP-11 POWER RESERVE 14 DAYS SAPPHIRE   REF. 911.JX.0102.RW

Movement: manual-winding HUB9011 Manufacture caliber; 14-day power reserve.
Functions: hours, minutes; power reserve indicator.
Case: polished sapphire crystal.
Dial: sapphire crystal.
Strap: white lined structured rubber.
Suggested price: $105,000

TECHFRAME FERRARI TOURBILLON CHRONOGRAPH CARBON YELLOW   REF. 408.QU.0129.RX

Movement: manual-winding HUB6311 caliber.
Functions: hours, minutes; tourbillon at 7; chronograph: 30-minute counter at 11, 60-second counter at 3.
Case: peek carbon fiber.
Dial: sapphire crystal and polished black.
Strap: black smooth rubber.
Suggested price: $137,000

TECHFRAME FERRARI TOURBILLON CHRONOGRAPH SAPPHIRE WHITE GOLD   REF. 408.JW.0123.RX

Movement: manual-winding HUB6311 Manufacture caliber.
Functions: hours, minutes; tourbillon at 7; chronograph: 30-minute counter at 11, 60-second counter at 3.
Case: microblasted 18K white gold.
Dial: sapphire crystal and gray opaline.
Strap: smooth gray rubber.
Suggested price: $149,000

BIG BANG UNICO GOLF   REF. 416.YS.1120.VR

Movement: automatic-winding HUB1580 UNICO Manufacture caliber.
Functions: hours, minutes; golf hole at 9; golf shot at 3; total score at 6.
Case: carbon fiber and gray Texalium upper layer.
Dial: matte black; satin-finished rhodium-plated appliques with white luminescent.
Strap: black rubber.
Suggested price: $31,500

HUBLOT

CLASSIC FUSION CHRONOGRAPH CERAMIC BLUE BRACELET — REF. 520.CM.7170.CM

Movement: automatic-winding HUB1143 caliber.
Functions: hours, minutes; small seconds at 3; date at 6; chronograph: central seconds hand, 30-minute counter at 9.
Case: satin-finished and polished black ceramic.
Dial: satin-finished blue sunray.
Bracelet: polished and satin-finished black ceramic.
Suggested price: $14,600

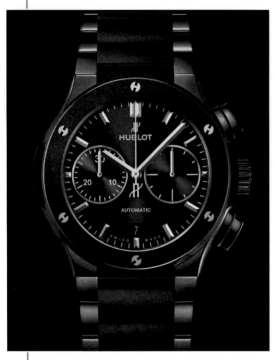

CLASSIC FUSION AEROFUSION CHRONOGRAPH ORLINSKI ALL BLACK — REF. 525.CI.0119.RX.ORL18

Movement: automatic-winding HUB1155 caliber.
Functions: hours, minutes; small seconds at 3; date at 6; chronograph: central seconds hand, 30-minute counter at 9.
Case: microblasted black ceramic; facetted design by Richard Orlinski.
Dial: sapphire crystal.
Strap: black smooth rubber.
Suggested price: $18,800

BIG BANG FERRARI UNICO MAGIC GOLD — REF. 402.MX.0138.WR

Movement: automatic-winding HUB1241 UNICO Manufacture caliber.
Functions: hours, minutes; date at 3; chronograph: central seconds hand, 60-minute counter at 3.
Case: polished 18K Magic Gold.
Dial: sapphire crystal.
Strap: black rubber and alcantara with red stitching.
Suggested price: $36,700

BIG BANG SHEPARD FAIREY — REF. 414.YF.1137.VR.SHF18

Movement: manual-winding HUB1201 Manufacture caliber; Ø 34.8mm, thickness: 6.8mm; 10-day power reserve; 223 components; 24 jewels; 21,600 vph.
Functions: hours, minutes; small seconds at 9; date at 3.
Case: carbon fiber and light texalium upper layer; Ø 45mm, thickness: 15.95mm; gray décor designed by Shepard Fairey; antireflective sapphire crystal; water resistant to 10atm.
Dial: skeletonized matte black.
Strap: black rubber and gray calfskin with embossed design by Shepard Fairey; titanium deployant buckle clasp.
Suggested price: $28,300

HUBLOT

MP-09 TOURBILLON BI-AXIS 3D CARBON — REF. 909.QD.1120.RX

Movement: manual-winding HUB9009.H1.RA Manufacture caliber; 5-day power reserve.
Functions: hours, minutes; date around hours and minutes dial; 5-day power reserve indicator at 9.
Case: 3D carbon fiber.
Dial: skeletonized.
Strap: black structured lined rubber.
Suggested price: $169,000

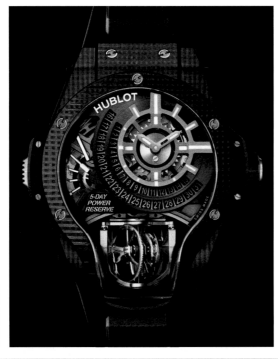

SPIRIT OF BIG BANG SAPPHIRE — REF. 601.JX.0120.RT

Movement: automatic-winding HUB4700 caliber.
Functions: hours, minutes; small seconds at 9; date between 4 and 5; chronograph: central seconds hand, 30-minute counter at 3, 12-hour counter at 6.
Case: polished sapphire crystal.
Dial: sapphire crystal.
Strap: transparent structured lined rubber.
Suggested price: $79,000

BIG BANG MECA-10 CERAMIC BLUE — REF. 414.EX.5123.RX

Movement: manual-winding HUB1201 Manufacture caliber.
Functions: hours, minutes
Case: microblasted and polished blue ceramic.
Dial: skeletonized matte blue.
Strap: black and blue structured lined rubber.
Suggested price: $22,000

BIG BANG UNICO TEAK ITALIA INDEPENDENT — REF. 411.OQ.1189.NR.ITI18

Movement: automatic-winding HUB1242 UNICO Manufacture caliber; 72-hour power reserve; 330 components; 38 jewels; 28,800 vph.
Functions: hours, minutes; date at 3; small seconds at 9; chronograph: central seconds hand, 60-minute counter at 3.
Case: microblasted and polished King Gold; Ø 45mm, thickness: 15.45mm; antireflective sapphire crystal; water resistant to 10atm.
Dial: skeletonized matte black.
Strap: black and orange structured lined rubber.
Suggested price: $40,900

IWC
SCHAFFHAUSEN

PERPETUAL POSSIBILITIES

The Portugieser collection is practically synonymous with IWC: its **GENEROUS PROPORTIONS, RAILWAY-STYLE MINUTE TRACK, APPLIED ARABIC NUMERALS, AND BEAUTIFULLY ORGANIZED DIAL HAVE CARRIED THE LINE'S STYLE SINCE THE 1930S**, even as the movements inside have grown ever more sophisticated. The Swiss watchmaker's new manufacturing facilities confirm the brand's philosophy and identity.

The Portugieser Perpetual Calendar is justly famous for its "double-moon" display, which indicates the phase of the moon in both the Northern and Southern Hemispheres simultaneously, with a small notation on the dial to specify which is which. Exceptionally accurate, the moonphase display deviates from actuality by just one day after 577.5 years—not an issue its current wearer will ever need to consider. The perpetual calendar mechanism needs no correction until March 1, 2100. At 3 o'clock, the date display shares a subdial with a display of the watch's considerable seven-day power reserve, with the months at 6 o'clock and the days and small seconds at 9 o'clock. A quite distinctive feature, the four-digit year display at 7:30 also deserves mention as well. Long a mainstay of the Portugieser collection for its attractive architecture and impressive mechanics, the Perpetual Calendar is available for the first time in a platinum case. At the center of this orchestra of indications beats the IWC 52615 caliber, conducting with aplomb. A solid 18-karat red-gold oscillating weight winds the movement automatically; its Pellaton winding also boasts durable ceramic components. The movement, complete with haute horology embellishments, is visible through the sapphire crystal caseback.

IWC

Long a mainstay of the Portugieser collection for its attractive architecture and impressive mechanics, the Perpetual Calendar is available for the first time in a platinum case.

The Portugieser Automatic showcases the collection's signature traits in a refined design. Its stainless steel case complements the silver-plated dial and rhodium hands, lending a monochrome sheen to the understated affair. Within beats the self-winding IWC 52010 movement, crafted in-house and bearing an 18-karat gold disc fitted into the winding rotor. The dial's streamlined aesthetic makes room for useful information, such as a power reserve display at 3 o'clock, date at 6 o'clock, and small seconds at 9 o'clock. The Portugieser's signature elegance guides the look of the Automatic, from the slim feuille hands to the simple Arabic numerals and the railway track chapter ring. A black alligator leather strap provides the perfect complement to this study in refinement. The impeccably finished caliber is visible through the sapphire crystal caseback.

Both watches—and the rest of IWC's production—were produced in IWC's Manufakturzentrum, just outside the Swiss town of Schaffhausen. This is where the brand creates its watches in accordance with the principles of its founders, Florentine Ariosto Jones, who believed in pairing traditional watchmaking with cutting-edge production techniques. Watchmakers begin with the raw materials and create individual components and assemble the finished product, all under the same roof. Though the movement components are largely produced by machine, it takes the finesse of the human hand to assemble them into the delicate marvel that is the modern watch.

▶ **PORTUGIESER AUTOMATIC**
The silver-plated dial, rhodium hands, and stainless steel case of the Portugieser Automatic lend it a cohesive monochrome elegance.

◀ **PORTUGIESER PERPETUAL CALENDAR**
The distinctive double moonphase and four-digit year display place the Portugieser Perpetual Calendar in a class of its own.

IWC

BIG PILOT'S WATCH — REF. IW501001

Movement: automatic-winding IWC 52110 Manufacture caliber; 168-hour power reserve; 31 jewels; 28,800 vph.
Functions: hours, minutes; 7-day power reserve indicator at 3; date at 6.
Case: stainless steel; Ø 46.2mm, thickness: 15.6mm; antireflective sapphire crystal; screw-down crown; water resistant to 6 bar.
Dial: black.
Strap: black calfskin leather.

BIG PILOT'S WATCH EDITION 'LE PETIT PRINCE' — REF. IW501002

Movement: automatic-winding IWC 52110 Manufacture caliber; 168-hour power reserve; 31 jewels; 28,800 vph.
Functions: hours, minutes; seconds; 7-day power reserve indicator at 3; date at 6.
Case: stainless steel; Ø 46.2mm, thickness: 15.6mm; screw-down crown; antireflective sapphire crystal; water resistant to 6 bar.
Dial: blue.
Strap: brown calfskin leather.

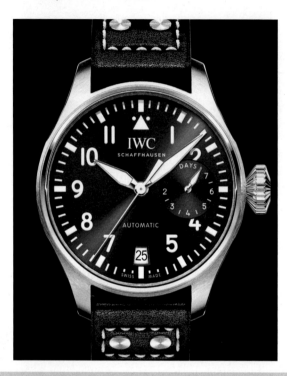

PILOT CHRONOGRAPH — REF. IW377709

Movement: automatic-winding 79320 caliber; 44-hour power reserve; 25 jewels; 28,800 vph.
Functions: hours, minutes; 60-second counter at 9; chronograph: 30-minute counter at 12, 12-hour counter at 6; day-date at 3.
Case: stainless steel; Ø 43mm, thickness: 15.5mm; antireflective sapphire crystal; screw-down crown; water resistant to 6 bar.
Dial: black.
Strap: black calfskin leather.

PILOT CHRONOGRAPH EDITION 'LE PETIT PRINCE' — REF. IW377714

Movement: automatic-winding 79320 caliber; 44-hour power reserve; 25 jewels; 28,800 vph.
Functions: hours, minutes; seconds at 9; chronograph: 30-minute counter at 12, 12-hour counter at 6; day-date at 3.
Case: stainless steel; Ø 43mm, thickness: 15.2mm; screw-down crown; antireflective sapphire crystal; water resistant to 6 bar.
Dial: blue.
Strap: brown calfskin leather.

IWC

PORTOFINO CHRONOGRAPH — REF. IW391009

Movement: automatic-winding IWC 75320 Manufacture caliber; 44-hour power reserve; 25 jewels; 28,800 vph.
Functions: hours, minutes; seconds at 9; chronograph: 30-minute counter at 12, 12-hour counter at 6; day-date at 3.
Case: stainless steel; Ø 42mm, thickness: 13.6mm; antireflective sapphire crystal; sapphire crystal; water resistant to 3 bar.
Dial: silver.
Bracelet: Milanese stainless steel mesh.

PORTOFINO HAND-WOUND 8 DAYS — REF. IW510103

Movement: manual-winding IWC 59210 Manufacture caliber; 192-hour power reserve; 30 jewels; 28,800 vph.
Functions: hours, minutes; seconds at 6; power reserve indicator at 9; date at 3.
Case: stainless steel; Ø 45mm, thickness: 11.7mm; antireflective sapphire crystal; sapphire crystal caseback; water resistant to 3 bar.
Dial: silver.
Strap: dark brown alligator leather.

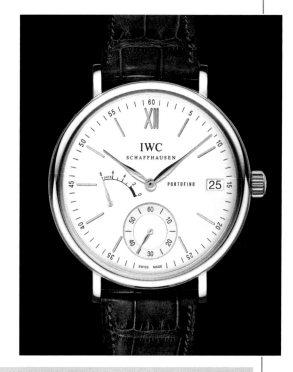

PORTOFINO HAND-WOUND 8 DAYS — REF. IW510115

Movement: manual-winding IWC 59210 Manufacture caliber; 192-hour power reserve; 30 jewels; 28,800 vph.
Functions: hours, minutes; seconds at 6; 8-day power reserve indicator at 9; date at 3.
Case: stainless steel; Ø 45mm, thickness: 11.7mm; sapphire crystal; sapphire crystal caseback; water resistant to 3 bar.
Dial: slate-gray dial.
Strap: gray suede.

PORTOFINO HAND-WOUND MOONPHASE — REF. IW516401

Movement: manual-winding IWC 59800 Manufacture caliber; 192-hour power reserve; 30 jewels; 28,800 vph.
Functions: hours, minutes; 60-second counter at 6; power reserve indicator at 9; date at 3; moonphase at 12.
Case: stainless steel; Ø 45mm, thickness: 13.2mm; antireflective sapphire crystal; sapphire crystal caseback; water resistant to 3 bar.
Dial: silver.
Strap: dark brown alligator leather.

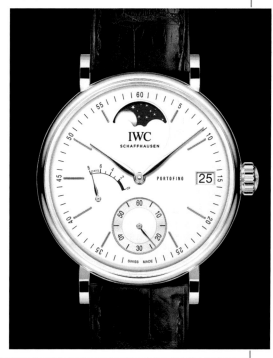

IWC

PORTUGIESER ANNUAL CALENDAR   REF. IW503502

Movement: automatic-winding IWC 52850 Manufacture caliber; 168-hour power reserve; 36 jewels; 28,800 vph.
Functions: hours, minutes, seconds; annual calendar: month, date, and day at 12; power reserve indicator at 3.
Case: stainless steel; Ø 44.2mm, thickness: 14.9mm; antireflective sapphire crystal; sapphire crystal caseback; water resistant to 3 bar.
Dial: blue.
Strap: black alligator leather.

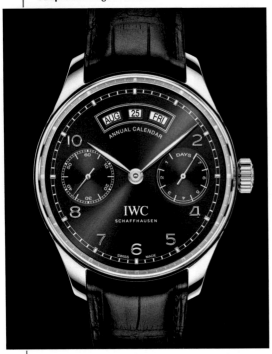

PORTUGIESER AUTOMATIC   REF. IW500701

Movement: automatic-winding IWC 52010 Manufacture caliber; 168-hour power reserve; 31 jewels; 28,8000 vph.
Functions: hours, minutes, seconds; 7-day power reserve indicator at 3; date at 6.
Case: 18K red gold; Ø 42.3mm, thickness: 14.2mm; sapphire crystal; sapphire crystal caseback; water resistant to 5 bar.
Dial: silver.
Strap: brown alligator leather.

PORTUGIESER CHRONOGRAPH   REF. IW371445

Movement: automatic-winding 79350 caliber; 44-hour power reserve; 31-jewels; 28,800 vph.
Functions: hours, minutes; chronograph: 60-minute counter at 6, 30-second counter at 12.
Case: stainless steel; Ø 40.9mm, thickness: 12.6mm; antireflective sapphire crystal; water resistant to 3 bar.
Dial: silver.
Strap: black alligator leather.

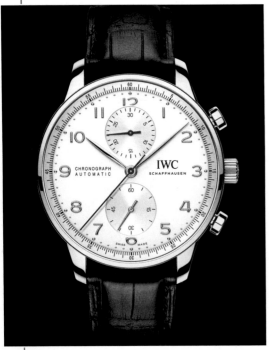

PORTUGIESER CHRONOGRAPH   REF. IW371491

Movement: automatic-winding 79350 caliber; 44-hour power reserve; 31 jewels; 28,800 vph.
Functions: hours, minutes; chronograph: 60-second counter at 6, 30-minute counter at 12.
Case: stainless steel; Ø 40.9mm, thickness: 12.6mm; antireflective sapphire crystal; sapphire crystal caseback; water resistant to 3 bar.
Dial: blue.
Strap: black alligator leather.

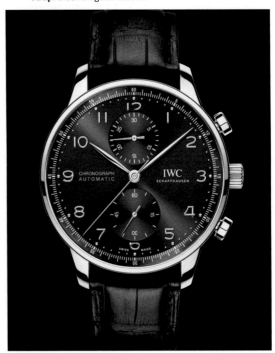

IWC

PORTUGIESER CHRONOGRAPH CLASSIC — REF. IW390302

Movement: automatic-winding IWC 89361 Manufacture caliber; 68-hour power reserve; 38 jewels; 28,800 vph.
Functions: hours, minutes, seconds; chronograph: hours and minutes at 12 o'clock, flyback; date at 3.
Case: stainless steel; Ø 42mm, thickness: 14.2mm; antireflective sapphire crystal; sapphire crystal caseback; water resistant to 3 bar.
Dial: silver.
Strap: black alligator leather.

PORTUGIESER PERPETUAL CALENDAR — REF. IW503401

Movement: automatic-winding IWC 52615 Manufacture caliber; 168-hour power reserve; 54 jewels; 28,800 vph.
Functions: hours, minutes; seconds at 9; 168-hour power reserve indicator at 3; moonphase at 12; perpetual calendar: month display at 6, day at 9, date at 7:30.
Case: 18K white gold; Ø 44.2mm, thickness: 14.9mm; antireflective sapphire crystal; sapphire crystal caseback; water resistant to 5 bar.
Dial: midnight blue.
Strap: black alligator leather.

PORTUGIESER PERPETUAL CALENDAR — REF. IW503406

Movement: automatic-winding IWC 52615 Manufacture caliber; 168-hour power reserve; 54 jewels; 28,800 vph.
Functions: hours, minutes; seconds at 9; 168-hour power reserve indicator at 3; moonphase at 12; perpetual calendar: month display at 6, day at 9, date at 3.
Case: platinum; Ø 44.2mm, thickness: 14.9mm; antireflective sapphire crystal; sapphire crystal caseback; water resistant to 3 bar.
Dial: silver.
Strap: black alligator leather.

PORTUGIESER YACHT CLUB CHRONOGRAPH — REF. IW390502

Movement: automatic-winding IWC 89361 Manufacture caliber; 68-hour power reserve; 38 jewels; 28,800 vph.
Functions: hours, minutes, seconds; chronograph: hours and minutes at 12, flyback; date at 3.
Case: stainless steel; Ø 43.5mm, thickness: 14.2mm; antireflective sapphire crystal; sapphire crystal caseback; water resistant to 6 bar.
Dial: silver.
Strap: black rubber.

A watchmaking, Leo Messi is widely considered one of the best players in the sport's history. To celebrate his once-in-a-lifetime skills, Jacob & Co. has created a watch in his honor: the Epic X Chrono Messi. The original Epic X Chronograph pioneered the collection's distinctive case shape, with elements that cross over onto the dial. Rugged and robust, the chronograph is water resistant to 200m and guaranteed shock resistant, with a skeleton bi-compax movement visible through a transparent caseback. This limited edition translates the Epic X look into titanium and white ceramic, which, in combination with the blue mineral crystal dial, recall the blue and white color scheme of the flag of Argentina, Messi's homeland. Tying it even more closely to its namesake superstar, the Epic X Chrono Messi bears the athlete's iconic number 10 in red on the flange ring, his logo on the dial at six o'clock, and his signature on the caseback. "I am delighted to be in partnership with Jacob & Co., whom I regard as the most innovative watchmaker," said Messi. This model, available in a limited edition of 180 pieces, is the first of several limited editions in a new three-year partnership between the athlete and the watchmaker.

JACOB & CO.

An 18-karat rose-gold plate decorated with The Godfather logo gives a visual hint to the theme of this musical watch, as do a miniature "godfather" and black lacquered piano at the center of the movement.

Continuing Jacob & Co.'s history of accomplishment in musical watches, the Opera Collection welcomes a new model that will appeal to fans of cinema and those fascinated by the world glimpsed in Francis Ford Coppola's towering Godfather trilogy. A pair of music-box cylinders plays a melody of 120 notes, evoking the emotional power of the beloved films. An 18-karat rose-gold plate decorated with The Godfather logo gives a visual hint to the theme, as do a miniature "godfather" and black lacquered piano at the center of the movement. The tourbillon rotates on no fewer than three axes: one rotation takes 24 seconds, another takes 8 seconds, and a third takes 30 seconds. The entire dial configuration rotates over 120° in 30 seconds, with the time display simultaneously rotating in the opposite direction to maintain its orientation vis-à-vis the wearer. Instead of a crown, the watch boasts an 18-karat rose-gold crank handle at 3 o'clock on the case for winding, and 18-karat rose-gold lift-out bows on the caseback for time-setting. The exclusive manual-winding manufacture Jacob&Co JCFM04 caliber powers this extraordinary machine, beating at a frequency of 21,600 vph and providing a power reserve of 50 hours. The matte black finish of the dial sets off the rose-gold accents and decorative elements.

◄ **EPIC X CHRONO MESSI** (*facing page*)
Combining the iconic look and robustness of the Epic X collection with touches that allude to Leo Messi's illustrious career, this 47mm limited edition model marks the beginning of a fruitful collaboration between the two virtuosos in their fields.

▲ **OPERA GODFATHER**
Two cylinders interact with two combs on the dial of the Opera Godfather to play a familiar tune, sharing the space with a fascinating triple-axis tourbillon.

JACOB & CO.

ASTRONOMIA SKY — REF. AT110.30.AA.SD.A

Movement: exclusive manual-winding Jacob & Co. JCAM11 caliber; Ø 40mm, thickness: 17.15mm; 60-hour power reserve; 395 components; 42 jewels; 21,600 vph; material: titanium; triple-axis gravitational tourbillon system; plate and bridges: hand-angled and polished, flank-drawn, circular-graining, polished sink; polished screws.
Functions: hours, minutes; sidereal time display: celestial panorama, oval sky indicator, month indicator; legal time display: day and night indicator: lacquered hand-engraved titanium globe inside a tinted half-domed sapphire; gravitational triple-axis tourbillon: rotating on 3 axes in 60 seconds, 5 minutes and on the central axis; orbital second indicator: rotating on 2 axes in 60 seconds and on the central axis, openworked titanium wheel and hand indicating seconds; hour and minute indicator: patented differential system, rotating on the central axis; Jacob Cut red moon rotating on 2 axes in 60 seconds and on the central axis (1 carat, 288 facets).
Case: 18K white gold; Ø 47mm, thickness: 25mm; unique domed antireflective sapphire crystal; sapphire crystal apertures on sides; winding and time-setting via 18K white-gold lift-out rotating "bows" on caseback; 18K white-gold caseback; water resistant to 3atm.
Dial: blue grade-5 titanium; 18K gold hand-engraved and applied stars and constellation; titanium hours and minutes subdial: hand-angled and polished, lacquered indexes, finished blue hands.
Strap: black alligator leather; 18K white-gold folding buckle.
Note: limited edition of 18 pieces.
Suggested price: $580,000
Also available: 18K rose-gold case.

ASTRONOMIA SOLAR — REF. AS310.40.SP.ZK.A

Movement: exclusive Manufacture manual-winding Jacob & Co. JCAM19 caliber; Ø 34.55mm, thickness: 16.5mm (11.70mm excluding citrine); 48-hour power reserve; 447 components; 43 jewels; 28,800 vph; material: titanium; double-axis gravitational tourbillon; finishing: plate and bridges: sand-blasted and angled; gear train: circular-graining, polished sink; polished screws; circular-graining barrel; flat balance spring.
Functions: hours; minutes dial rotating in 10 minutes on the central axis; patented differential gears system allows it to keep the 12/6 position; rose-gold and blue lacquered globe rotating in 60 seconds on 2 axes; flying tourbillon cage in 60 seconds; citrine Jacob-Cut® moon (1.5 carats, 288 facets); frame carrying 3 Jacob-Cut® gemstones: amethyst, topaz and peridot; entire movement rotates clockwise 360° in 10 minutes; aventurine base rotates counter-clockwise 360° in 10 minutes carrying 7 planets made of cabochon-cut precious stone half-spheres: white granite for Mercury, rhodonite for Venus, red jasper for Mars, pietersite for Jupiter, tiger eye for Saturn (18K rose-gold appliqué for the rings), blue calcite for Uranus and lapis lazuli for Neptune.
Case: 18K rose gold and sapphire; Ø 43.40mm, thickness: 21mm; unique domed antireflective sapphire crystal; winding and time-setting via two 18K rose-gold "bows"; sapphire crystal caseband; aventurine sky layer front case; 18K rose-gold caseback; water resistant to 3atm.
Dial: blue PVD; rose-gold-plated engraving of Zodiac signs; rose-gold hands.
Strap: alligator leather; 18K rose-gold folding clasp.
Note: individually numbered limited edition.
Suggested price: $318,000

ASTRONOMIA OCTOPUS BAGUETTE — REF. AT802.40.BD.UA.A

Movement: exclusive Manufacture manual-winding Jacob & Co. JCAM24 caliber; Ø 40mm, thickness: 17.15mm; 60-hour power reserve; 367 components; 42 jewels; 21,600 vph; vertical movement with 4 arms; material: titanium; sapphire barrel bridge; double-axis gravitational tourbillon; finishing: plate and bridges: hand angled and polished, flank draw, circular-graining, polished sink, one sapphire bridge; polished screws; barrel: circular graining; conical pinions; 4 mechanical ball bearing devices; flat balance spring.
Functions: hours, minutes; double-axis gravitational tourbillon, 1st cage rotating over 60 seconds and 2nd cage over 2.5 minutes; 1-carat exclusive Jacob-Cut® diamond: 288 facets, rotating on itself in 30 seconds; magnesium blue lacquered globe, rotating on itself in 30 seconds; hours and minutes dial, patented differential gears system allows to keep the 12/6 position; top 3D octopus hand-engraved in titanium, hand-painted and lacquered rotating on the central axis (weight: 2.90g).
Case: 18K rose gold; Ø 50mm, thickness: 25.45mm; set with 281 baguette-cut white diamonds (approx. 9.89 cts); lugs set with 80 baguette-cut white diamonds (approx. 6.10 cts); sapphire apertures on side; unique domed sapphire crystal with anti-reflective treatment; sapphire caseback with anti-reflective treatment gold metallization; winding and time-setting via 18K rose-gold lift-out rotating "bows" at caseback; water resistant to 30m.
Dial: titanium; hand angled and polished; lacquered indexes; gun-blue finish hands.
Strap: alligator leather; 18k rose-gold folding buckle.
Note: unique piece.
Price: available upon request.

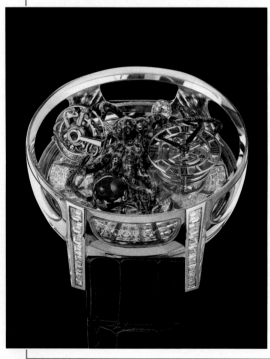

TWIN TURBO TOURBILLON MINUTE REPEATER — REF. TT100.40.NS.NK.A

Movement: exclusive manual-winding JCFM01 caliber; Ø 34.43mm, thickness: 13.2mm; 72-hour power reserve; 572 components; 49 jewels; 21,600 vph; twin triple-axis tourbillon system with decimal minute repeater; hand-angled and polished plate and bridges; flank-drawn, circular-graining; circular barrel and plate; polished screw; pinions: conical, 4 mechanical ball bearing devices; 2 screw balances.
Functions: hours, minutes; decimal minute repeater; cathedral gong; twin triple-axis tourbillon: 40 seconds, 8 minutes and 3 minutes; regulator between the two triple-axis tourbillon carriages; double mechanical safety features during a chiming sequence.
Case: 18K rose gold; 57.3x51mm, thickness: 16.9mm; domed sapphire crystal; domed sapphire crystal caseback; water resistant to 3atm.
Dial: smoked sapphire crystal with applied "Jacob&Co." logo; red neoralithe inner ring with second graduation; black skeletonized hands with SuperLumiNova coating.
Strap: black alligator leather; 18K rose-gold deployment buckle.
Note: limited edition of 18 pieces.
Suggested price: $395,000
Also available: carbon and 18K white-gold case; carbon and grade-5 titanium case with black DLC treatment.

JACOB & CO.

EPIC X CHRONO ROSE GOLD   REF. EX300.40.SR.RR.B

Movement: exclusive automatic-winding Jacob & Co. JCAA05 caliber; Ø 30.4mm; 48-hour power reserve; 260 components; 27 jewels; 28,800 vph; column-wheel balance screws system; anti-shock; column-wheel finishes: angled, polished and drawn; anthracite rotor with red lacquered engraving "Jacob & Co. Genève", circular-graining bridges.
Functions: hours, minutes; small seconds at 9; chronograph: central seconds hand, 30-minute counter at 3.
Case: 18K rose gold; Ø 47mm; thickness: 14mm; white ceramic bezel; antireflective sapphire crystal; satin-finished and polished edges; red mineral crystal and sapphire crystal caseback; water resistant to 20atm.
Dial: satin-finished red-gold anodized aluminum inner rings; skeletonized hands with white SuperLumiNova; red chronograph hands.
Strap: white honeycomb rubber; 18K rose-gold and titanium deployment clasp.
Suggested price: $33,000

OPERA BY JACOB & CO. – GODFATHER MUSICAL WATCH   REF. OP110.40.AG.AB.A

Movement: exclusive Manufacture manual-winding Jacob & Co. JCFM04; Ø 43mm, thickness: 17.2mm; 50-hour power reserve; 658 components; 58 jewels; 21,600 vph; material: steel, brass, platinum and titanium; triple-axis tourbillon; music box with double combs; finishing: plates and bridges: shot-blasted and black PVD finish; circular-graining barrels; rose-gold-plated cylinders; angled-bevel and mirror-polished screws.
Functions: off-centered hours and minutes; triple-axis gravitational tourbillon cage rotating on 3 axes in 24 seconds, 8 seconds and 30 seconds; Godfather melody activated by pushbutton at 10: rotation of 2 cylinders against 2 combs decorated by 18K rose-gold Godfather plate and black lacquered piano; miniature Godfather at the center of the movement: rotation of 120° in 30 seconds.
Case: 18K rose gold and sapphire; Ø 49mm, thickness: 23mm; unique domed antireflective sapphire crystal; winding and time-setting via two 18K rose-gold "bows" in the caseback; 18K rose-gold caseback; water resistant to 3atm.
Dial: polished matte black; 18K rose-gold appliqués and indexes; finished blue hands.
Strap: black alligator leather; 18K rose-gold folding clasp.
Note: limited edition of 18 pieces.
Suggested price: $360,000

EPIC SF 24 TOURBILLON   REF. ES102.24.NS.LC.A

Movement: exclusive automatic-winding Jacob & Co. JCAA03 caliber; 35.5x42.1mm, thickness: 12.2mm; 48-hour power reserve; 528 components; 36 jewels; 28,800 vph; glucydur balance wheel system; anti-shock; high-end finishing; sand-blasted.
Functions: hours, minutes; small seconds on tourbillon at 10; 24-hour world time indicator at 12.
Case: black DLC grade-5 titanium; Ø 45mm, thickness: 13.65mm at the center, 16.7mm at the world time display; satin-finished, micro-blasted and polished; antireflective sapphire crystal; water resistant to 3atm.
Dial: anthracite opaline; gold skeletonized central hands with white SuperLumiNova coating.
Strap: black alligator leather; black DLC grade-5 titanium deployment buckle.
Note: limited edition of 101 pieces.
Suggested price: $175,000
Also available: set with baguette diamonds; 18K rose-gold case.

MILLIONAIRE YELLOW DIAMOND   REF. ML510.50.YD.AA.A40B

Movement: exclusive manual-winding Jacob & Co. JCAM23 caliber; 32.7x25.2mm; 72-hour power reserve; 197 components; 19 jewels; 21,600 vph; hand-angled, sand-blasted, flank-drawn.
Functions: hours, minutes.
Case: 18K white gold; 46x35mm, thickness: 11mm; set with 54 Asscher-cut yellow diamonds; antireflective sapphire crystal; crown set with 1 rose-cut yellow diamond; sapphire crystal caseback; water resistant to 3atm.
Dial: indexes set with 11 baguette diamonds; gun blue skeletonized hands.
Bracelet: 18K white gold set with 203 Asscher-cut yellow diamonds.
Note: individually numbered pieces.
Price: available upon request.

JACOB & CO.

CAVIAR TOURBILLON DIAMOND BRACELET   REF. CV300.40.BD.BD.A

Movement: exclusive manual-winding Jacob & Co. JCBM05 caliber; Ø 30.4mm, thickness: 9.1mm; 100-hour power reserve; 169 components; 19 jewels; 21,600 vph; titanium balance with gold timing screw; anti-shock; finishing: bridges set with 455 brilliant-cut white diamonds (~1.22 carats).
Functions: hours, minutes.
Case: polished 18K rose gold; Ø 44mm, thickness: 15.85mm; set with 200 baguette white diamonds (~20.59 carats); crown set with 14 baguette white diamonds and one rose-cut white diamond (~1.01 carats); antireflective sapphire crystal; water resistant to 3atm.
Dial: 123 baguette white diamonds (~7.72 carats); skeletonized hands.
Bracelet: 18K rose gold; set with 464 baguette white diamonds (~49.34 carats).
Note: limited edition of 18 pieces.
Suggested price: $1,000,000

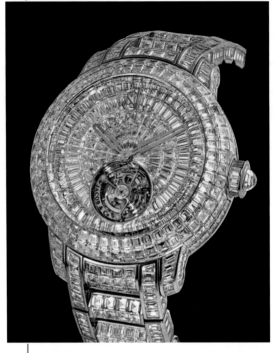

BRILLIANT FLYING TOURBILLON – ICE CREAM   REF.

Movement: manual-winding Jacob&Co. JCBM01a caliber; Ø 34.1mm, thickness: 6mm; 100-hour power reserve; 206 components; 29 jewels; 21,600 vph; titanium balance system with gold timing screw; anti-shock; circular satin-brushed "Jacob & Co." upper bridge of the tourbillon cage; satin-finished, hand-polished angles, drawn-finished flanks; bridges set with 346 brilliant-cut white diamonds (~1.5 carats).
Functions: hours, minutes; tourbillon at 6; power reserve indicator.
Case: 18K rose gold; Ø 47mm, thickness: 14.65mm; invisibly set with 170 Ice Cream-cut pink sapphires (approx. 27.63 cts); crown set with 14 Ice Cream-cut pink sapphires and one cabochon-cut pink sapphire (approx. 1.18 cts); antireflective sapphire crystal; circular satin-finished sapphire crystal caseback with polished gadroon power reserve indicator; water resistant to 3atm.
Dial: invisibly set with 146 Ice Cream-cut pink sapphires (approx. 14.9 cts); leaf-shaped hands.
Strap: white alligator leather; 18K rose-gold tang buckle set with 21 Ice Cream-cut pink sapphires (approx. 2.53 cts).
Note: set with total of 351 Ice Cream-cut pink sapphires and 1 cabochon-cut pink sapphire (approx. 46.24 cts); limited edition of 18 pieces.
Price: available upon request.

BRILLIANT FULL BAGUETTE AUTOMATIC   REF. BA534.40.BR.BR.A

Movement: Swiss automatic-winding JCAA04 caliber; Ø 25.6mm; 21 jewels; 28,800 vph.
Functions: hours, minutes.
Case: polished 18K rose gold; Ø 44mm, thickness: 12mm; invisibly set with 226 baguette rubies (~16.31 carats); crown set with 12 baguette rubies and one rose-cut ruby (~0.84 carat); antireflective sapphire crystal; circular satin-finished and engraved caseback.
Dial: invisibly set with 187 baguette rubies (~12.06 carats); leaf-shaped hands.
Strap: green alligator leather; 18K rose-gold tang buckle set with 21 baguette rubies (~1.43 carats).
Note: individually numbered pieces.
Suggested price: $500,000
Also available: satin strap.

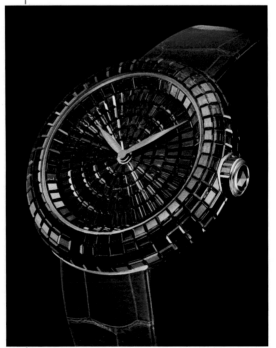

BRILLIANT MYSTERY BAGUETTE   REF. BM526.30.BE.BE.A

Movement: Swiss quartz; Jacob & Co. JCGQ063 caliber; Ø 11mm; 5 jewels.
Functions: hours, minutes.
Case: polished 18K white gold; Ø 38mm; invisibly set with 291 baguette emeralds (~16.89 carats); antireflective sapphire crystal; time setting by a corrector; water resistant to 3atm.
Dial: invisibly set with 51 natural gemstones; two floating black triangle hands.
Strap: red satin; 18K white-gold tang buckle set with 28 baguette diamonds or natural gemstones.
Note: limited edition of 18 individually numbered pieces.
Suggested price: $770,000

JACOB & CO.

BRILLIANT ART DECO BLUE SAPPHIRE   REF. BT545.30.BB.BB.A

Movement: exclusive manual-winding Jacob & Co. JCAM05 caliber; Ø 32.6mm, thickness: 6mm; 72-hour power reserve; 172 components; 19 jewels; 21,600 vph; openworked movement with sapphire plates and finely-set bridges: 171 brilliant-cut white diamonds (~0.56 carat) and one brilliant-cut blue sapphire (~0.11 carat); hand-polished angles.
Functions: hours, minutes; tourbillon at 6.
Case: 18K white gold; Ø 47mm, thickness: 14.35mm; polished and invisibly set with 205 baguette blue sapphires (~21.7 carats); inside upper bezel with 48 baguette white diamonds (~2.11 carats); crown set with 14 baguette blue sapphires (~0.77 carat) and one rose-cut blue sapphire (~0.52 carat); antireflective sapphire crystal; water resistant to 3atm.
Dial: skeletonized; gun blue leaf-shaped hands.
Strap: blue alligator leather; 18K white-gold tang buckle set with 21 baguette blue sapphires (~1.71 carats).
Note: unique piece.
Suggested price: $750,000

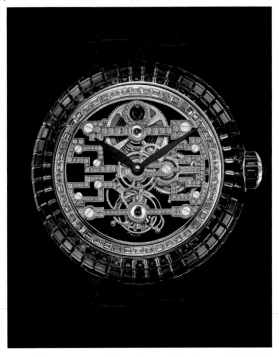

BRILLIANT POCKET WATCH PENDANT   REF. BS200.40.RD.CB.A

Movement: exclusive manual-winding Jacob & Co. JCAM01 caliber; Ø 32mm, thickness: 5.8mm; 46-hour power reserve; 150 components; 21 jewels; 21,600 vph; angled bridges with sand-blasted finishing, angled steel elements with drawn-finishing; openworked anthracite movement with rose-gold finishing with "Jacob & Co." logo, bridges and gears.
Functions: hours, minutes.
Case: 18K rose gold; Ø 42.5mm, thickness: 13.8mm; set with one row of 72 brilliant-cut diamonds (~0.68 carat); crown set with 31 brilliant-cut diamonds (~0.39 carat); antireflective sapphire crystal; engraved sapphire crystal caseback; water resistant to 3atm.
Dial: skeletonized; leaf-shaped hands.
Chain: 18K rose gold.
Note: limited edition of 101 pieces.
Suggested price: $69,000

EPIC X ROSE GOLD   REF. EX100.40.PS.RW.A

Movement: exclusive manual-winding Jacob & Co. JCAM02 caliber; Ø 14.25mm, thickness: 5.9mm; 48-hour power reserve; 158 components; 21 jewels; 28,800 vph; barrel and balance wheel vertically aligned; barrel with sliding clamp system; balance screws visible on the front side; time-setting spring with three functions: ratchet, lever and setting lever; anti-shock; "Jacob & Co." upper bridge; sand-blasted, angled and drawn-finished bridges; anthracite and red lacquered "Epic X" engraving on the right-side bridge; mirror-polished finishing screws; circular-grained bridges on the caseback.
Functions: hours, minutes.
Case: 18K rose gold; Ø 44mm, thickness: 12.3mm; satin-finished and polished edges; engraving in the inner ring; antireflective sapphire crystal; sapphire crystal caseback; water resistant to 5atm.
Dial: red neoralithe inner ring; skeletonized hands.
Strap: black open-worked honeycomb rubber; 18K rose-gold tang buckle.
Note: individually numbered pieces.
Suggested price: $18,000
Also available: alligator leather strap.

ASTRONOMIA TABLECLOCK   REF. AT900.10.AC.MT.A

Movement: manual-winding Jacob&Co. JCAM17 caliber; 103x54.5mm; 8-day power reserve; material: titanium; 399 components; 41 jewels; 21,000 vph; finishes: sand-blasted, diamond angled, circular graining; Côtes de Genève.
Functions: hours, minutes; rotating dial over 10 minutes on the central axis with patented differential gears system; aluminum lacquered Earth globe rotating in 60 seconds on 2 axes; mechanical gravitational rotating wheels on 2 axes: 1st axis: over 2.5 minutes, 2nd axis: over 60 seconds; genuine meteorite stone rotating over 60 seconds.
Case: stainless steel; Ø 128mm, thickness: 85mm; mineral crystal; winding and time-setting via stainless steel key on the back; aventurine and stainless steel sky layer front case unique domed mineral crystal with anti-reflective treatment.
Dial: titanium; gun-blue-finished hands.
Note: individually numbered limited edition of 101 pieces.
Suggested price: $160,000

JD
JAQUET DROZ

BEYOND PERFECTION

Returning to its favorite themes, Jaquet Droz further refines and clarifies the emblems of its brand, **DEVELOPING NEW CREATIONS THAT TAKE EACH CONCEPT EVEN FURTHER INTO ITS IDENTITY**.

A signature feature of Jaquet Droz's collections, the Grande Seconde focuses the attention on the often-overlooked seconds display, lavishing care on a display that reminds us to live in the moment. The new Grande Seconde Skelet-One Ceramic rethinks the concept in several important ways, affirming the watchmaker's commitment to the details of haute horology. A case diameter of 41.5mm, paired with a sapphire crystal "glass box" front and back, provides a generous stage for the automatic-winding Jaquet Droz 2663 SQ movement within, a completely skeletonized masterpiece that transforms the machinery into a vision of lightness. The movement plate extends to the very edge of the case, replacing the casing ring for a design in which every superfluous element has been pared away. The cutting-edge case is forged in black ceramic, for a masculine look that perfectly complements the movement's black NAC coating, the visible blued screws, and blue canvas strap.

◀ **GRANDE SECONDE SKELET-ONE CERAMIC**
Every element of the Grande Seconde Skelet-One Ceramic contributes to its goal of transparency, from its minimal bridges to the entirely skeletonized 18-karat gold oscillating weight.

JAQUET DROZ

No fewer than seven animations spring into action on the Tropical Bird Repeater's astonishing dial, including a world-first hummingbird whose wings beat up to 40 times per second.

Stunning its public with its exceptionally detailed, superlatively lovely automaton pieces, Jaquet Droz takes us on a journey to the rain forest, where incredibly diverse flora and fauna coexist and compete in a lush landscape. Engravers and miniature painters bring to life a living tableau, animated by the watchmaker's renowned automatons. No fewer than seven animations spring into action on this astonishing dial, including a world-first hummingbird whose wings beat up to 40 times per second. In addition to the animations, some of which last up to 12 seconds, the Tropical Bird Repeater features the "queen of complications," a minute repeater sounded by cathedral gongs within the case. The innovative Jaquet Droz RMA89 drives this feast for the eyes and ears, beating at 21,600 vph and providing a newly enhanced power reserve of 60 hours. A special timepiece released in a limited edition of just 8 pieces, the Tropical Bird Repeater is mounted on a dark green alligator leather strap.

Proving that marvelous things come in small packages, Jaquet Droz applies its signature figure 8 to the Lady 8 Petite, a diminutive offering that shrinks the dimensions of the Lady 8 to 25mm in diameter while losing nothing of its appeal. A simple hours and minutes display dominates the dial, crafted in semi-precious material. Graduated diamonds swirl along the sides of the case, sweeping up and turning to embrace a stone mounted above 12 o'clock on the dial: a pearl, rounded aventurine, or other material to match the dial. The self-winding movement within boasts a 38-hour power reserve, as well as a guilloché oscillating weight with a fanned design. The fashionably feminine touches on these stainless steel models extend to the strap, a slim double-wrap model in black or blue grained calfskin, adorned with a diamond-set clasp.

▲ TROPICAL BIRD REPEATER
Expert in both minute repeaters and automatons, Jaquet Droz presents a stunning handcrafted dial with a preening peacock, fluttering dragonflies, and a sonnerie that chimes out the hours, quarter-hours, and minutes.

◀ LADY 8 PETITE
Jaquet Droz combines its favored figure 8 with beautiful feminine touches on a 25mm steel case, including a pearl or aventurine ball that rolls under the touch of a finger.

JAQUET DROZ

GRANDE SECONDE MOON BLACK ENAMEL — REF. J007533201

Movement: automatic-winding Jaquet Droz 2660QL3; silicon balance spring and pallet horns; double barrel; flat bridges; 18K red-gold oscillating weight; 30 jewels; 68-hour power reserve; 28,800 vph.
Functions: off-centered hours and minutes; large off-centered seconds; pointer-type date display at 6; astronomical moonphase at 6; hours, minutes, seconds and date hands in 18K red gold; date with red varnished tip.
Case: 18K red gold; Ø 43mm, thickness: 13.23mm; individual serial number engraved on the caseback; water resistant to 3 bar.
Dial: double-level black Grand Feu enamel; 18K red-gold applied ring; black onyx moon disc; star and moon appliqués in 18K red gold and 22K gold respectively.
Strap: rolled-edge handmade black alligator leather; 18K red-gold ardillon buckle.
Suggested price: $28,900

GRANDE SECONDE SKELET-ONE RED GOLD — REF. J003523240

Movement: automatic-winding Jaquet Droz 2663 SQ; skeletonized; black treatment, silicon balance spring; pallet horns; double barrel; 18K red-gold oscillating weight; individual number engraved on the oscillating weight; 30 jewels; 68-hour power reserve; 28,800 vph.
Functions: off-centered hours and minutes; large off-centered seconds.
Case: 18K red gold; Ø 41mm, thickness: 12.30 mm; water resistant to 3 bar.
Dial: sapphire; hour circle in 18K white gold; fixing screws in 18K red gold; 18K red-gold hands.
Strap: rolled-edge handmade black alligator leather; 18K red-gold ardillon buckle.
Suggested price: $33,600

LADY 8 PETITE JADE — REF. J014603171

Movement: automatic-winding Jaquet Droz 615; silicon balance spring and pallet horns; 18K red-gold oscillating weight, 29 jewels; 38-hour power reserve; 21,600 vph.
Functions: hours, minutes.
Case: 18K red gold; set with 68 diamonds (0.52 ct) Ø 25mm, thickness: 10.23mm; individual limited edition serial number engraved on the weight; jade bead at 12; jade crown cabochon; water resistant to 3 bar.
Dial: copper-colored mother-of-pearl; circular grain motif.
Bracelet: 18K red gold; set with 192 diamonds (2.31 cts) with 18K red-gold folding clasp set with 21 diamonds (0.14 ct).
Note: limited edition of 8 pieces.
Suggested price: $92,400

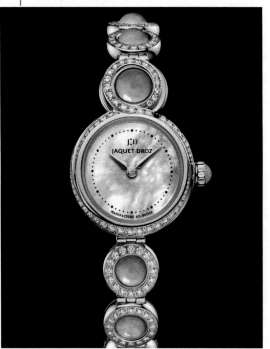

GRANDE SECONDE OFF-CENTERED ONYX — REF. J006010270

Movement: automatic-winding Jaquet Droz 2663.P; silicon balance spring and pallet horns; double barrel; flat bridges; heavy metal oscillating weight; 30 jewels; 68-hour power reserve; 28,800 vph.
Functions: off-centered hours and minutes; large off-centered seconds.
Case: stainless steel; Ø 39mm, thickness: 12.62mm; winding-stem at 4; individual serial number engraved on the caseback; water resistant to 3 bar.
Dial: black onyx; 18K white-gold applied ring; rhodium-treated hands.
Strap: rolled-edge handmade black alligator leather; stainless steel folding clasp.
Suggested price: $10,700

JAQUET DROZ

GRANDE SECONDE QUANTIEME SILVER — REF. J007030242

Movement: automatic-winding Jaquet Droz 2660Q2; double barrel; heavy metal oscillating weight; 30 jewels; 68-hour power reserve; 28,800 vph.
Functions: off-centered hours and minutes; large seconds; pointer-type date display at 6.
Case: stainless steel; Ø 43mm, thickness: 11.63mm; individual serial number engraved on the caseback; water resistant to 3 bar.
Dial: silver opaline; applied ring; hours, minutes, seconds and date hands in blued steel; date hand with red varnished tip.
Strap: rolled-edge handmade black alligator leather; stainless steel folding clasp.
Suggested price: $9,100

PETITE HEURE MINUTE SMALTA CLARA TIGER — REF. J005503500

Movement: automatic-winding Jaquet Droz 6150; silicon balance spring and pallet horns; platinum oscillating weight with 18K red-gold applique; 29 jewels; 38-hour power reserve; 21,600 vph.
Functions: off-centered hours and minutes.
Case: 18K red gold; set with 100 diamonds (0.89 ct); Ø 35mm, thickness: 10.85mm; water resistant to 3 bar.
Dial: 18K white-gold with "plique-à-jour" Grand Feu enamel; white mother-of-pearl subdial; 18K red-gold hands.
Strap: rolled-edge handmade brown satin.
Note: limited edition of 28 pieces; individual limited edition serial number engraved on the oscillating weight.
Suggested price: $54,600

GRANDE HEURE MINUTE ARDOISE — REF. J017030240

Movement: automatic-winding Jaquet Droz 1169.Si; silicon balance spring; pallet horns; double barrel; oscillating weight in heavy metal; 31 jewels; 68-hour power reserve; 28,800 vph.
Functions: hours, minutes; small seconds at 9.
Case: stainless steel; Ø 43mm, thickness: 11.77mm; individual serial number engraved on the caseback.
Dial: slate-gray; applied ring with rhodium treatment.
Strap: rolled-edge handmade slate-gray alligator leather; stainless steel ardillon buckle.
Suggested price: $15,200

LOVING BUTTERFLY AUTOMATON AVENTURINE — REF. J032534271

Movement: automatic-winding Jaquet Droz 2653 AT1; silicon balance spring and pallet horns; double barrel; flat bridges; 22K white-gold oscillating weight; hand-winding automaton with mechanism for moving the butterfly's wings and the wheel of the chariot; crown pushbutton triggering mechanism; triple barrel.
Functions: off-centered hours and minutes; automatons.
Case: 18K white gold; set with 212 diamonds (0.68 ct); Ø 43mm, thickness: 16.63mm; individual limited serial number engraved on the caseback.
Dial: aventurine; 18K white-gold decorations; white mother-of-pearl hour circle; 18K white-gold appliqués of cherub, chariot and butterfly body; butterfly wings in 18K white gold; chariot wheel in 18K white gold.
Strap: rolled-edge handmade blue alligator leather; 18K white-gold folding clasp set with 32 diamonds (0.26 ct).
Note: limited edition of 28 pieces.
Suggested price: $136,000

LONGINES

LA GRANDE CLASSIQUE DE LONGINES ADOPTS THE BRAND'S ICONIC BLUE COLOR

La Grande Classique de Longines has played a major role in forging the reputation of the winged hourglass brand, now bearing the Swiss watchmaker's signature blue color for dials and straps. A symbol of the classic elegance of Longines, the line, launched in 1992, has welcomed a number of new variations over the years. With this new version in blue, La Grande Classique de Longines preserves its timeless refinement and style as it adapts to modern tastes.

The collection that is emblematic of the winged hourglass brand, La Grande Classique de Longines has been gaining male and female fans since its debut in 1992. Over the years, the line has expanded and offered new versions, without ever compromising its classic spirit and elegance. Longines now introduces its iconic blue to the dials and straps of this collection.

This new interpretation is sure to appeal to those who appreciate modern refinement with this new color, while maintaining its timeless style. The heart of this collection is the delicate slim profile of La Grande Classique de Longines, which is made possible by the unique construction of these elegant timepieces. The back of the case also serves as the lugs for the strap, a technique that is patented by Longines.

Whether inlaid with diamonds or presenting a simpler elegance, the new models in this iconic collection come in four sizes (24mm, 29mm, 36mm and 37mm) to fit perfectly on both male and female wrists. La Grande Classique de Longines with blue sunray dial is available with inlaid stone, inlaid diamond or Roman numeral hour markers, with either a blue alligator strap that matches the dial or a stainless steel bracelet.

LONGINES

COSC
CONTRÔLE OFFICIEL SUISSE
DES CHRONOMÈTRES

RECORD
— COLLECTION —
CERTIFIED CHRONOMETER – SILICON BALANCE-SPRING

THE LONGINES RECORD COLLECTION GOES FOR GOLD

In the purest watchmaking traditions of Longines, the watches in the Record collection combine classic elegance with excellence, aspiring to become the spearheads of the brand. These exceptional timepieces, whose movement includes a silicon balance spring featuring unique properties. Longines is now introducing new variations in rose gold or in stainless steel and rose gold, certified as "chronometers" by the Swiss Official Chronometer Testing Institute (COSC).

Through its Record collection, Longines expresses its very essence, combining timeless elegance with uncompromising excellence, a guaranteed recipe for success, representing a bridge between tradition and innovation. Longines now offers new models adorned with rose gold to further enrich this resolutely classic collection,—the epitome of the expertise of the Swiss watchmaker—making them appeal to men and women with discerning taste in watches.

The Record collection features two separate and distinct sized calibers and in order to enhance the accuracy and longevity of its timepieces, Longines integrates silicon balance springs into its movements. This lightweight material is resistant to corrosion, and is unaffected by normal temperature variations, magnetic fields, and atmospheric pressure.

Such excellence deserves recognition. Thanks to their extreme accuracy, the watches in the Record collection have been awarded "chronometer" certification by the COSC (Contrôle officiel suisse des chronomètres—Swiss Official Chronometer Testing Institute). Equipped with automatic movements, all of the timepieces were tested separately by this neutral and independent organization, and are authorized to have the designation "CHRONOMETER" stamped on the dial. Bestowed with a high degree of added value, these certified "chronometers" now join the ranks of truly exceptional timepieces.

The watches of the Record collection are available in four sizes (26, 30mm, 38.5mm and 40mm) and a variety of dials, and are intended for men or women. The stainless steel and rose gold or 18-karat rose-gold case features a transparent back that allows for admiration of the detail of the movement. Some models also offer a version inlaid with diamonds. The watch is available with a stainless steel bracelet with a rose-gold cap, or with a black, brown or blue alligator strap matched to the dial and equipped with a security folding clasp.

THE LONGINES MASTER COLLECTION

AN ANNUAL CALENDAR FOR THE LONGINES MASTER COLLECTION: A DEMONSTRATION OF THE WATCHMAKING EXPERTISE OF LONGINES

As a traditional watchmaking company, Longines has been producing exceptional timepieces since the very beginning. The Longines Master Collection is the perfect contemporary illustration of this concept, as demonstrated by the success of this range since its launch in 2005. Fitted with mechanical movements, the models in this collection offer many special features, which now includes a timepiece with an annual calendar—a guaranteed winner for those who appreciate exceptional watchmaking.

In 2005, Longines launched the line that would become its flagship: The Longines Master Collection. From the start, this collection has enjoyed an undeniable level of success propelling its status as the iconic representation of Longines watchmaking expertise. Over the years, The Longines Master Collection has been enhanced by new features, but has never lost sight of the timeless classical elegance that is emblematic of the brand, and that contributes to its global success. Longines now offers an annual calendar as part of its Master Collection, controlled by a new mechanical movement, exclusive to Longines.

The annual calendar automatically manages the varying lengths of the months. This allows the timepieces to distinguish between a month with 30 days and a month with 31 days, with no manual intervention required. The Longines Master Collection is the first range from Longines to house this feature. A true bargain for those who value timepieces with an exceptional quality-price ratio.

Combining classical elegance with excellence, the models from The Longines Master Collection featuring an annual calendar contain the new automatic L897 caliber in a stainless steel case measuring 40mm in diameter. With 64 hours of power reserve and the annual calendar displayed at 3 o'clock, the movement can be viewed through the transparent caseback. The rhodium-plated or blued steel hands jump off the black or silver barleycorn or sunray blue dial. These elegant timepieces are complimented by a steel bracelet or black, brown or blue alligator watch strap, all fitted with a folding safety clasp.

LONGINES

LONGINES PRIMALUNA — REF. L8.112.0.87.6

Movement: quartz caliber L129.
Functions: hours, minutes, seconds; date at 3.
Case: stainless steel; Ø 30mm; set with 48 Top Wesselton VVS diamonds (0.403 ct); scratch-resistant sapphire crystal with several layers of anti-reflective coating on the underside; water resistant to 3 bar.
Dial: white mother-of-pearl; 11 diamond hour markers; blued steel hands.
Bracelet: stainless steel with triple safety folding clasp and pushpieces.
Price: available upon request.

LONGINES DOLCEVITA — REF. L5.512.5.79.7

Movement: quartz caliber L176.
Functions: hours, minutes; small seconds at 6.
Case: stainless steel with a solid rose-gold crown; 23.30x37.00mm; rectangular; set with 46 Top Wesselton VVS diamonds (0.552 ct); scratch-resistant sapphire crystal; water resistant to 3 bar.
Dial: silver "flinqué"; painted Roman numerals; blued steel hands.
Bracelet: stainless steel and rose gold cap 200 with triple safety folding clasp and pushpiece opening mechanism.
Price: available upon request.

LONGINES SYMPHONETTE — REF. L2.305.5.87.7

Movement: quartz caliber L963.
Functions: hours, minutes.
Case: stainless steel with a solid rose-gold crown; 18.90x29.40mm; oval; scratch-resistant sapphire crystal with several layers of anti-reflective coating on the underside; water resistant to 3 bar.
Dial: white mother-of-pearl; 13 diamond hour markers; pink hands.
Bracelet: stainless steel and rose-gold cap 200 with triple safety folding clasp and pushpiece opening mechanism.
Price: available upon request.

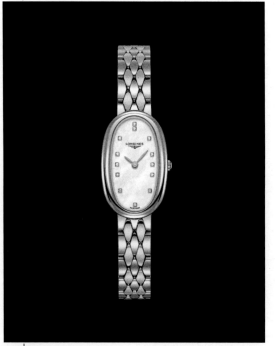

LA GRANDE CLASSIQUE DE LONGINES — REF. L4.512.4.94.6

Movement: quartz caliber L420.
Functions: hours, minutes.
Case: stainless steel; Ø 29mm; scratch-resistant sapphire crystal; water resistant to 3 bar.
Dial: sunray blue, painted Roman numerals; silvered polished hands.
Bracelet: stainless steel with triple safety folding clasp and pushpiece opening mechanism.
Price: available upon request.

LONGINES

THE LONGINES ELEGANT COLLECTION — REF. L4.910.4.92.2

Movement: automatic-winding caliber L619/L888; 11½ lines; 64-hour power reserve; 21 jewels; 25,200 vph.
Functions: hours, minutes, seconds; date at 3.
Case: stainless steel; Ø 39mm; scratch-resistant sapphire crystal; transparent caseback with sapphire crystal; water resistant to 3 bar.
Dial: sunray blue; applied indexes; silvered polished hands.
Strap: blue alligator leather; with buckle.
Price: available upon request.

THE LONGINES 1832 — REF. L4.825.4.92.2

Movement: automatic-winding caliber L888; 11½ lines; 64-hour power reserve; 21 jewels; 25,200 vph.
Functions: hours, minutes, seconds; date at 3.
Case: stainless steel; Ø 40mm; scratch-resistant sapphire crystal with several layers of anti-reflective coating on the underside; transparent caseback with sapphire crystal; water resistant to 3 bar.
Dial: beige; applied indexes and dots in SuperLumiNova®; silvered polished hands with SuperLumiNova®.
Strap: brown alligator leather; with buckle.
Price: available upon request.

THE LONGINES MASTER COLLECTION — REF. L.2.910.4.78.3

Movement: automatic-winding caliber L897; 11½ lines; 64-hour power reserve; 21 jewels; 25,200 vph.
Functions: hours, minutes, seconds; month and date by annual calendar at 3.
Case: stainless steel; Ø 40mm; scratch-resistant sapphire crystal; transparent caseback with sapphire crystal; water resistant to 3 bar.
Dial: silver; "barleycorn" motif; painted Arabic numerals; blued steel hands.
Strap: brown alligator leather; with triple safety folding clasp and pushpiece opening mechanism.
Price: available upon request.

CONQUEST CLASSIC — REF. L2.786.4.56.6

Movement: automatic-winding caliber L688 with a column-wheel chronograph mechanism; 13¼ lines; 54-hour power reserve; 27 jewels; 28,800 vph.
Functions: hours, minutes; small seconds at 9; date at 4; chronograph: 30-minute counter at 3, 12-hour counter at 6, central 60-second hand.
Case: stainless steel; Ø 41mm; scratch-resistant sapphire crystal with several layers of anti-reflective coating on the underside; screw-down and transparent caseback with sapphire crystal; water resistant to 5bar.
Dial: black; applied Arabic numeral and indexes with SuperLumiNova®; silvered polished hands with SuperLumiNova®.
Bracelet: stainless steel with triple safety folding clasp and pushpieces.
Price: available upon request.

LONGINES

RECORD COLLECTION — REF. L2.820.8.92.2

Movement: automatic-winding caliber L888.4; 11½ lines; 64-hour power reserve; 21 jewels; 25,200 vph; chronometer certified by the COSC.
Functions: hours, minutes, seconds; date at 3.
Case: 18K rose gold; Ø 38.50mm; scratch-resistant sapphire crystal with several layers of anti-reflective coating on the underside; transparent caseback with sapphire crystal; water resistant to 3 bar.
Dial: sunray blue, applied indexes; pink hands.
Strap: blue alligator leather; with triple safety folding clasp and pushpieces.
Price: available upon request.

EQUESTRIAN — REF. L6.141.4.77.6

Movement: quartz caliber L178.
Functions: hours, minutes; small seconds at 6.
Case: stainless steel; rectangular; 22x32mm; scratch-resistant sapphire crystal with several layers of anti-reflective coating on the underside; water resistant to 3 bar.
Dial: silvered-colored "checkerboard"; 12 diamond indexes; blued steel hands.
Bracelet: stainless steel; with triple safety folding clasp and pushpieces.
Price: available upon request.

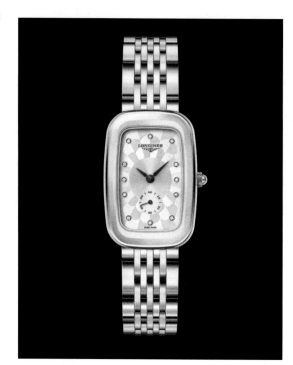

HYDROCONQUEST — REF. L3.781.4.96.6

Movement: automatic-winding caliber L619/888; 11½ lines; 64-hour power reserve; 21 jewels; 25,200 vph.
Functions: hours, minutes, seconds; date at 3.
Case: stainless steel; Ø 41mm; scratch-resistant sapphire crystal with several layers of anti-reflective coating on both sides; screw-down case back; screw-in crown; unidirectional rotating bezel with ceramic; water resistant to 30 bar.
Dial: sunray blue, applied Arabic numerals and indexes with SuperLumiNova®; silvered polished hands with SuperLumiNova®.
Bracelet: stainless steel; with double folding clasp and integrated diving extension.
Price: available upon request.

CONQUEST — REF. L3.380.4.76.6

Movement: quartz caliber L296.
Functions: hours, minutes, seconds; moonphase display at 6; date at 3.
Case: stainless steel; Ø 29.50mm; scratch-resistant sapphire crystal with several layers of anti-reflective coating on the underside; screw-down caseback; screw-in crown; water resistant to 30 bar.
Dial: sunray silver; applied Arabic numerals and indexes; silvered polished hands.
Bracelet: stainless steel; with triple safety folding clasp and pushpieces.
Price: available upon request.

LONGINES

CONQUEST V.H.P.   REF. L3.726.4.76.6

Movement: quartz caliber L288.
Functions: hours, minutes, seconds; date (perpetual calendar) at 3.
Case: stainless steel; Ø 43mm; scratch-resistant sapphire crystal with several layers of anti-reflective coating on the underside; screw-down case back; water resistant to 5 bar.
Dial: silver; applied Arabic numerals and indexes with SuperLumiNova®; black hands with SuperLumiNova®.
Bracelet: stainless steel; with triple safety folding clasp and pushpieces.
Price: available upon request.

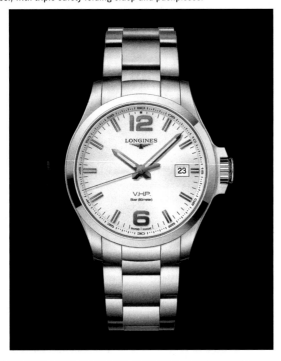

CONQUEST V.H.P. GMT FLASH SETTING   REF. L3.728.4.96.9

Movement: quartz caliber L287.
Functions: hours, minutes, seconds; 24-hour hand with time zone mechanism; date (perpetual calendar) at 3.
Case: stainless steel; Ø 43mm; scratch-resistant sapphire crystal with several layers of anti-reflective coating on the underside; screw-down caseback; water resistant to 5 bar.
Dial: blue; applied Arabic numerals and indexes with SuperLumiNova®; black hands with SuperLumiNova®.
Strap: blue rubber; with double safety folding clasp and pushpieces.
Price: available upon request.

THE LONGINES SKIN DIVER WATCH   REF. L2.822.4.56.6

Movement: automatic-winding caliber L/888; 11½ lines; 64-hour power reserve; 21 jewels; 25,200 vph.
Functions: hours, minutes, seconds.
Case: stainless steel; Ø 42mm; scratch-resistant sapphire crystal with several layers of anti-reflective coating on both sides; screw-down case back; screw-in crown; unidirectional rotating bezel with black PVD treatment; water resistant to 30 bar.
Dial: black; painted Arabic numerals and indexes with SuperLumiNova®; silvered polished hands with SuperLumiNova®.
Bracelet: stainless steel; with pushpiece opening mechanism.
Price: available upon request.

THE LONGINES HERITAGE MILITARY   REF. L2.819.4.93.2

Movement: automatic-winding caliber L888; 11½ lines; 64-hour power reserve; 21 jewels; 25,200 vph.
Functions: hours, minutes, seconds.
Case: stainless steel; Ø 38.50mm; scratch-resistant sapphire crystal with several layers of anti-reflective coating on the underside; water resistant to 3 bar.
Dial: silvered opalin with vintage spraying; painted Arabic numerals; blued steel hands.
Bracelet: green leather strap with buckle.
Price: available upon request.

LOUIS VUITTON
READY TO EXPLORE

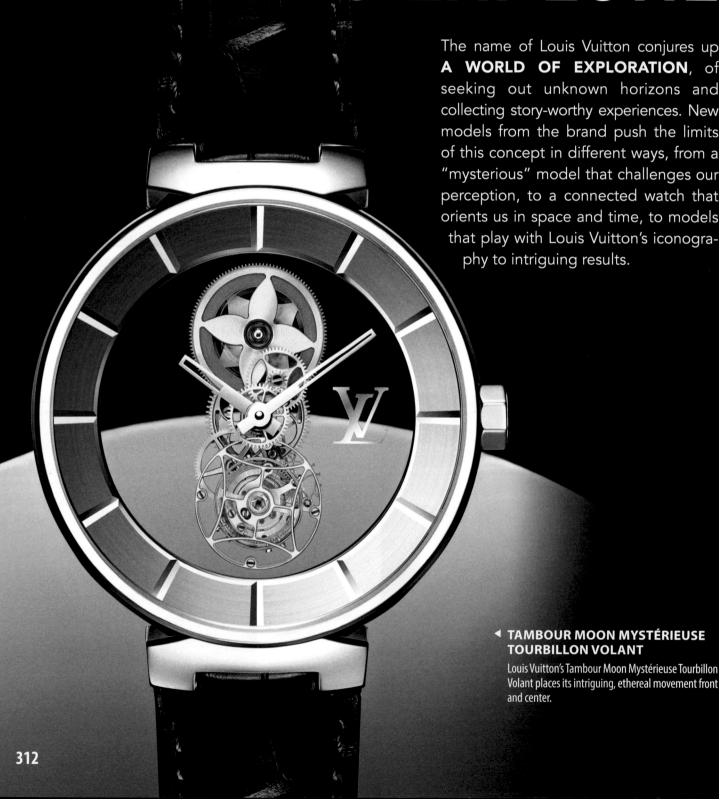

The name of Louis Vuitton conjures up **A WORLD OF EXPLORATION**, of seeking out unknown horizons and collecting story-worthy experiences. New models from the brand push the limits of this concept in different ways, from a "mysterious" model that challenges our perception, to a connected watch that orients us in space and time, to models that play with Louis Vuitton's iconography to intriguing results.

◄ **TAMBOUR MOON MYSTÉRIEUSE TOURBILLON VOLANT**

Louis Vuitton's Tambour Moon Mystérieuse Tourbillon Volant places its intriguing, ethereal movement front and center.

LOUIS VUITTON

The "mysterious" watch turns a sophisticated knowledge of horology into a magical experience for the viewer, as the watch hands appear to float in mid-air.

The horological craftsmanship of Louis Vuitton has been in intense development for some time, with complicated models such as the brand's Répétition Minutes, Skeleton Tourbillon Poinçon de Genève, and Spin Time showing the ingenuity of which it is capable. The Tambour Moon Mystérieuse Tourbillon Volant begins with the tourbillon, already a sign of incredible watchmaking sophistication for any high-end horologer. Creating a "flying" tourbillon, which fixes the tourbillon carriage at just one end, adds a layer of difficulty for the watchmaker, and pays off with a spectacular, weightless-seeming miniature marvel. La Fabrique du Temps Louis Vuitton, the brand's manufacture in Geneva, then combines this with a "mysterious" movement. The latter technique was first developed in the 19th century, turning a sophisticated knowledge of watch movements into a magical experience for the viewer. Through a canny understanding of where and how to manipulate movement components, the watch becomes an almost magical creation, with hands that appear to levitate unsupported. (In fact, they are integrated into transparent sapphire crystals.)

Louis Vuitton has adopted this entrancing technique for its manual-winding LV 110 caliber, blending it with classic LV style. A Monogram Flower shields the co-axial double barrel at 12 o'clock, and the entire movement descends from this foundation in a vertical arrangement. The central wheels for the hours and minutes appear to float just above the flying tourbillon carriage at 6 o'clock. The airy atmosphere is enhanced by the disconnect between the winding crown and double barrel, as well as the choice of a flying tourbillon. Another Monogram Flower, this one openworked, adorns the tourbillon carriage, which completes a full rotation every 60 seconds. In addition to a striking visual, the LV 110 also provides an eight-day power reserve. Complementing the minimalist beauty of the movement, the watch is housed within a 45mm concave platinum case and mounted on a black alligator strap. The owner of this marvel may choose to personalize it with his or her initials on the back of the tourbillon carriage.

LOUIS VUITTON

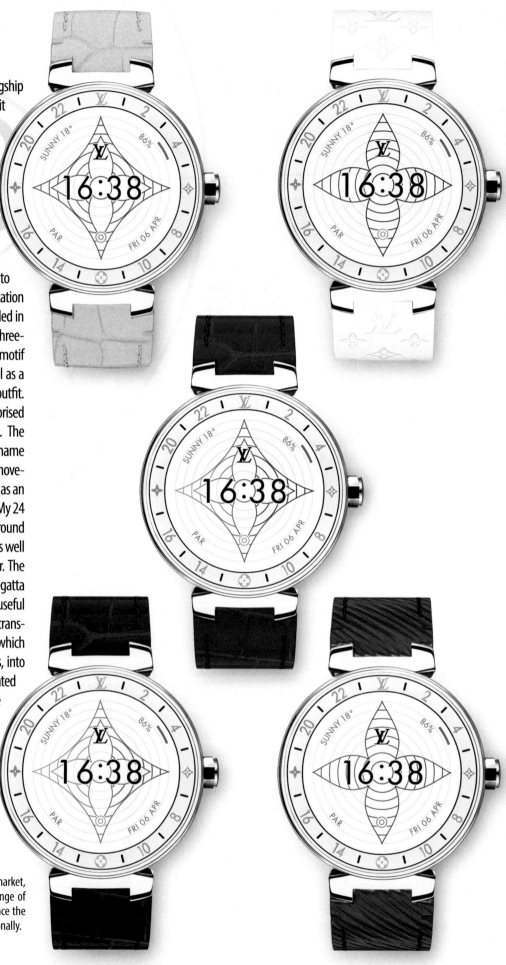

The Tambour Horizon is Louis Vuitton's flagship model in the connected watch realm, and it has continued to grow and change since its introduction in 2017, incorporating new features and opportunities for customization. A 24-hour ring around the dial uses rose-gold-colored accents to express the model's aesthetic vision and specific Louis Vuitton identity. The new option of a white ring around the dial lends a feminine touch to a versatile unisex look. Louis Vuitton stays true to its penchant for travel and quest for personalization with the dial designs and useful features included in this connected watch. The user can select the three-flower Monogram or checkerboard Damier motif as the main design element of the dial, as well as a choice of colors to complement any mood or outfit. The time display thus keeps the wearer apprised against an exclusive Louis Vuitton backdrop. The GMT function allows the wearer to display the name of any major city and its time with a single movement, and the model establishes its bona fides as an essential travel accessory with the option of a "My 24 Hours" dial, bedecked with Louis Vuitton flags around the hours, minutes, day, and date indications, as well as options for the weather and a step counter. The Regatta dial picks up the original Tambour Regatta watch's color scheme and design, as well as its useful stopwatch function, while the Spin Time dial translates that collection's innovative principle, in which rotating cubes indicate the hours and minutes, into pixels. Other dial options include colorful, animated Kabuki masks and Chinese Zodiac motifs. The practical City Guide feature acts as a personal tour guide, queuing up the top destinations and hidden gems for 29 cities around the globe on the Tambour Horizon's screen. A wide variety of strap colors provides the wearer with another aspect that they can customize to their exact specifications.

▶ **TAMBOUR HORIZON**

A formidable contender in the connected watch market, Louis Vuitton's Tambour Horizon offers a wide range of straps and customizable dial displays that reference the illustrious brand's history both visually and functionally.

LOUIS VUITTON

A streamlined variation on Louis Vuitton's iconic Tambour model, the two-tone All Black version has taken on chic rose-gold accents in its newest iteration. The Monogram Flower at the center of the dial is the most salient evidence of the warm approach, with simple hour markers and Arabic numerals radiating out, each one punctuated by a diamond on the deep flange for a hint of luxury. Adapting to the needs of its clientele, this model is available in three diameters: 28mm, 34mm, and 39.5mm, each version realized in stainless steel with glossy black PVD coating. A more imposing model, at 46mm, the Tambour Chronographe Automatique applies the style to a chronograph setup, with stunning results. The mechanical movement inside beats within a stainless steel case bearing a matte black PVD coating that underscores the chronograph's more serious demeanor. The subdials are arranged in a formation that echoes the large "V" for Vuitton that angles across the dial, an emblem of the company since Gaston-Louis Vuitton introduced it in 1901. Rose-gold touches spread evenly around the watch, from the chronograph pushbuttons to the lugs, to the hands, hour markers, and subdials, making for a seamlessly coherent whole. In keeping with Louis Vuitton's emphasis on personalization, all four models use the brand's patented system for switching between straps quickly and easily.

▼ **TAMBOUR ALL BLACK & GOLD CHRONOGRAPHE AUTOMATIQUE** *(below left)*
Louis Vuitton combines the chronograph complication with time-tested chic on its All Black & Gold Chronograph Automatique.

▼ **TAMBOUR ALL BLACK & GOLD** *(below right)*
The two-tone concept of the Tambour All Black & Gold lends a rose-gold warmth to the chic All Black concept.

LOUIS VUITTON

The history of Louis Vuitton luggage plastered with labels and decals from exotic locales provides a well of striking images to draw from for the Tambour World Tour. The dials of watches in this charming collection evoke the excitement of new places and the luxury of Louis Vuitton's lineage. The Tambour World Tour collection hearkens back to the days of luggage that told more of its owner's story with every sticker added to the exterior. The playful dials, developed by La Fabrique du Temps Louis Vuitton, each boast a backdrop that reproduces a classic Louis Vuitton canvas: either the Monogram canvas or the Damier Graphite motif. Evocative designs emulate the travel labels of days gone by. Even the straps, interchangeable via Louis Vuitton's patented system, take up the canvas patterns. The steel case comes in a diameter of 28mm, 34mm, 39.5mm, and 41.5mm, supplying the right size for any wrist. A quartz movement keeps perfect time within.

▼▶ **TAMBOUR WORLD TOUR WATCH**
Available in a variety of sizes and with an interchangeable strap, the Tambour World Tour references the history of Louis Vuitton and the romance of international travel.

LOUIS VUITTON

For those who appreciate the romance of travel in every aspect of their lives, the World Tour concept translates to a miniature steel clock as well. At 65mm in diameter, the spherical timepiece slips easily into one's suitcase to travel the world with its owner. A GMT function graces the Damier Cobalt dial, indicating home time with a miniature rocket ship on a 24-hour ring. A dark blue theme sets off the colorful sticker collection, with the shade woven into the dial and incorporated into the PVD treatment of the clock's base.

▲▶ **TAMBOUR WORLD TOUR CLOCK**
The charming Tambour World Time Clock measures a diminutive 65mm in diameter, making it easy to pack in matching luggage.

LOUIS VUITTON

TAMBOUR MOON MYSTERIEUSE FLYING TOURBILLON REF. Q8E810

Movement: manual-winding LV110 caliber; developed and assembled at la Fabrique du Temps Louis Vuitton.
Functions: hours, minutes; flying tourbillon at 6.
Case: platinum 950; Ø 45mm; water resistant to 5atm.
Dial: skeletonized with line of wheels.
Strap: black alligator leather; platinum 950 ardillon buckle.
Price: available upon request.

TAMBOUR WORLD TOUR DAMIER GRAPHITE REF. QA063Z + R15046

Movement: quartz.
Functions: hours, minutes, seconds.
Case: polished stainless steel; Ø 41.5mm; Damier pattern engraved caseback; water resistant to 10atm.
Dial: Damier graphite motifs and stickers; hands with SuperLumiNova.
Strap: Damier graphite and stickers interchangeable strap fitted with Louis Vuitton's patented system; stainless steel ardillon buckle.
Price: available upon request.

TAMBOUR ALL BLACK & GOLD CHRONOGRAPH REF. QA039Z + R17029

Movement: automatic-winding.
Functions: hours, minutes; small seconds at 3; date between 4 and 5; chronograph: 12-hour counter at 6, 30-minute counter at 9, central seconds hand.
Case: stainless steel with matte black PVD coating; Ø 46mm; 18K rose-gold horns, crown and pushpieces; sapphire crystal with LV logo on the caseback; water resistant to 10atm.
Dial: matte black with contrasting anthracite V; Gaston V signature at 12, pink-gold-colored hour markers and semi-skeletonized hands with SuperLumiNova.
Strap: coal black alligator leather Interchangeable strap fitted with Louis Vuitton's patented system; ardillon buckle with black DLC coating.
Price: available upon request.

TAMBOUR HORIZON MONOGRAM WHITE REF. QA000Z + R15139

Movement operating system: Wear OS by Google.
Functions: Louis Vuitton customizable watch faces; "My 24 Hours" watch face; City Guide and My flight dedicated travel application; iconic Louis Vuitton watch dials.
Case: polished stainless steel; Ø 42mm, thickness: 12.55mm; sapphire crystal with a 24-hour ring, numbers and golden Monogram flower motifs; sapphire glass caseback with Louis Vuitton logo; water resistant to 3atm.
Dial: My Digital Monogram; Multifunctional AMOLED touch screen, ultra high definition; resolution: 390x390.
Strap: Monogram rubber Interchangeable strap fitted with Louis Vuitton's patented system; stainless steel ardillon buckle.
Price: available upon request.

LOUIS VUITTON

TAMBOUR SLIM STAR BLOSSOM 28MM — REF. Q1K0CZ + R15192

Movement: quartz.
Functions: hours, minutes.
Case: 18K pink gold; Ø 28mm; gem-set bezel; caseback engraved with Monogram; LV engraved crown; water resistant to 5atm.
Dial: white mother-of-pearl with Monogram Flowers in pink gold and diamonds.
Strap: pink alligator leather interchangeable strap fitted with Louis Vuitton's patented system; 18K pink-gold ardillon buckle.
Price: available upon request.

TAMBOUR SLIM STAR BLOSSOM 35MM — REF. QA048Z + R17059

Movement: automatic-winding.
Functions: hours, minutes.
Case: 18K pink gold; Ø 35mm; gem-set bezel; caseback engraved with Monogram; LV engraved crown; water resistant to 5atm.
Dial: mother-of-pearl with Monogram Flowers in pink gold and diamonds.
Strap: pomegranate alligator leather interchangeable strap fitted with Louis Vuitton's patented system; 18K pink-gold ardillon buckle.
Price: available upon request.

TAMBOUR ALL BLACK 34MM — REF. QA048Z + R15167

Movement: quartz.
Functions: hours, minutes, seconds; date at 3.
Case: polished stainless steel with glossy black PVD coating; Ø 34mm; black sapphire crystal caseback with printed Monogram Flowers; water resistant to 10atm.
Dial: glossy black with golden Monogram Flower; diamond hour markers, black polished hour and minute luminescent hands.
Strap: black Monogram vernis interchangeable strap fitted with Louis Vuitton's patented system; stainless steel ardillon buckle.
Price: available upon request.

TAMBOUR WORLD TOUR MONOGRAM — REF. QA061Z + R15962

Movement: quartz.
Functions: hours, minutes, seconds.
Case: polished stainless steel; Ø 34mm; caseback engraved with Monogram Flowers pattern; water resistant to 10atm.
Dial: printed Monogram pattern with stickers; hands with SuperLumiNova.
Strap: Monogram canvas and stickers interchangeable strap fitted with Louis Vuitton's patented system; stainless steel ardillon buckle:
Price: available upon request.

OMEGA

FROM THE SEA TO THE STARS

Always changing, never content to rest on its laurels, Omega **REVISITS THREE ICONIC COLLECTIONS**, making alterations both inside and out for a selection that is more current and vital than ever. From deep beneath the ocean to the stars over a city skyline, Omega makes its mark.

Long a flagship of Omega's collections, the Seamaster Diver 300m has dominated the world beneath the waves for two and a half decades, as well as gracing the silver screen on the wrist of iconic spy Bond... James Bond. It returns with a design in which every detail has been rethought, taking nothing for granted in the quest for a cutting-edge watch in both technology and design. Fourteen new variations—six in stainless steel and eight in gold—build upon the foundation of this iconic model. The movement inside is the Master Chronometer Calibre 8800, an upgrade that enhances the watch's precision, performance, and magnetic resistance. The case design is paramount in a diving watch, and this one sports a new ceramic bezel with the diving scale in Ceragold™ (an exclusive, specialized alloy) or white enamel. The indexes are luminescent and raised, for greater visibility. Among other aesthetic changes, the date window has moved to 6 o'clock and the skeleton hands have undergone subtle but unmistakable tweaks. The watch helium escape valve, a crucial feature on any diving watch, has evolved into an ingenious patented device with a conical shape and the capacity for underwater operation.

◀ **OMEGA SEAMASTER DIVER 300M**
Subtle engraved waves on the dial of the Omega Seamaster Diver 300m nod to the collection's iconic status as Omega's flagship diving model.

◀ Daniel Craig wears the Omega Seamaster Diver 300m. (*far left*)

OMEGA

The Aqua Terra dial bears 191 diamonds of varying sizes set in undulating waves, interspersed with waves in silvery guilloché and Sedna™ gold.

The Seamaster collection has also revisited the exquisite Aqua Terra line, adding jewelry touches that delight and dazzle fans. The Sedna™ gold case sets the tone, bringing a warm blush to the feminine models and serving as a complementary frame for the 12 marquise-cut rubies on the dial. Brilliant-cut diamonds frame the dial, which bears 191 diamonds of varying sizes set in undulating waves, interspersed with waves in silvery guilloché and Sedna™ gold. The assortment of techniques perfectly captures the scintillating beauty of evening sunlight dancing on the sea. The glossy red leather strap completes the look with a fierce splash of color that highlights the ruby hour markers. The watch lives up to its name, with a 38mm case that provides water resistance to 150m.

New grace notes and aesthetic adjustments update the Constellation Manhattan collection, bringing the stately family solidly up to date. First officially launched in 1952, the Constellation collection has been a touchstone for Omega's feminine side ever since, and the Manhattan line, introduced in 1982, added a few trademark signs of sophistication. The perfectly round dial, tonneau-shaped case (complete with half-moon facets), and four "claws" that extend across the bezel immediately mark out the line's classic look. Each 29mm model in the collection contains the Calibre 8700/8701, which has undergone eight demanding tests to earn its designation as "Master Chronometer." Diamonds serve as hour markers on the face and sparkle thickly on the bezel, their glitter enhancing the lustrous glow of the mother-of-pearl dial and the chic two-tone case in Sedna™ gold and stainless steel. The diminutive star above the date at 6 o'clock alludes to the collection's history. The transparent caseback reveals the delicate finishing of the movement, which is rhodium-plated and bears côtes de Genève in a softly radiating swirl. With 101 new ladies' models in the collection, any woman can find her perfect match.

▲ **SEAMASTER AQUA TERRA**
This jeweled Seamaster Aqua Terra boasts beauty, sturdiness, and "brains," with the Master Chronometer Calibre 8807 beating within.

▶ **CONSTELLATION MANHATTAN**
The distinctive shape of the case and construction of the bracelet in the Constellation Manhattan collection lend each model a look of savvy sophistication.

OMEGA

SEAMASTER AQUA TERRA JEWELRY 38MM
REF. 220.58.38.20.99.005

Movement: automatic-winding Omega Caliber 8807 Certified Master Chronometer; 55-hour power reserve.
Functions: hours, minutes, seconds.
Case: Sedna™ gold; Ø 38mm; antireflective sapphire crystal; sapphire crystal caseback; water resistant to 15 bar.
Dial: white.
Strap: blue leather.

DEVILLE TRÉSOR QUARTZ 36MM
REF. 428.58.36.60.02.001

Movement: quartz Omega Caliber 4061; 48-month power reserve.
Functions: hours, minutes.
Case: Sedna™ gold; Ø 36mm; antireflective sapphire crystal; water resistant to 3 bar.
Dial: silver.
Strap: brown leather.

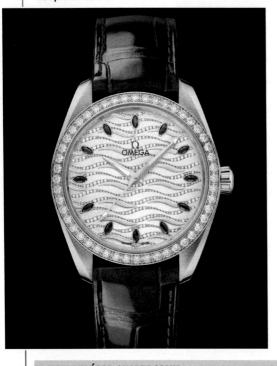

DEVILLE TRÉSOR QUARTZ 36MM
REF. 428.18.36.60.11.001

Movement: quartz Omega caliber 4061; 48-month power reserve.
Functions: hours, minutes.
Case: steel; Ø 36mm; antireflective sapphire crystal; water resistant to 3 bar.
Dial: red.
Strap: red leather.

DEVILLE TRÉSOR QUARTZ 36MM
REF. 428.17.36.60.04.001

Movement: quartz Omega Caliber 4061; 48-month power reserve.
Functions: hours, minutes.
Case: steel; Ø 36mm; antireflective sapphire crystal; water resistant to 3 bar.
Dial: white.
Strap: blue fabric.

OMEGA

CONSTELLATION MANHATTAN CO-AXIAL MASTER CHRONOMETER 29MM
REF. 131.55.29.20.55.001

Movement: automatic-winding Omega 8701 caliber Certified Master Chronometer; 50-hour power reserve.
Functions: hours, minutes, seconds; date.
Case: Sedna™ gold; Ø 29mm; antireflective sapphire crystal; sapphire crystal caseback; water resistant to 5 bar.
Dial: white.
Bracelet: gold.

CONSTELLATION MANHATTAN CO-AXIAL MASTER CHRONOMETER 29MM
REF. 131.25.29.20.55.002

Movement: automatic-winding Omega 8700 caliber Certified Master Chronometer; 50-hour power reserve.
Functions: hours, minutes, seconds; date.
Case: steel and yellow gold; Ø 29mm; antireflective sapphire crystal; sapphire crystal caseback; water resistant to 5 bars.
Dial: white.
Bracelet: steel and gold.

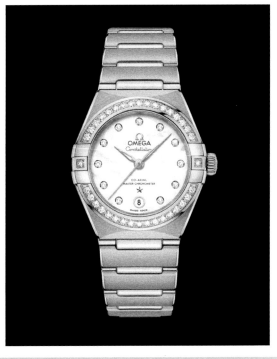
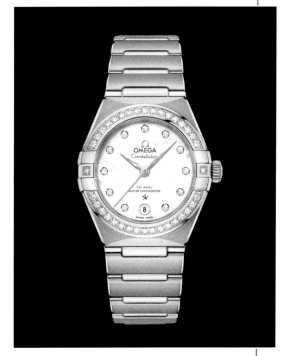

SPEEDMASTER 38 CO-AXIAL CHRONOGRAPH 38MM
REF. 324.28.38.50.02.001

Movement: automatic-winding Omega Caliber 3300; 52-hour power reserve.
Functions: hours, minutes, running seconds at 9; chronograph: 30-minute counter at 3, 12-hour counter at 6; date at 6.
Case: steel; Ø 38mm; antireflective sapphire crystal; sapphire crystal caseback; water resistant to 10 bar.
Dial: silver.
Strap: green leather.

SEAMASTER AQUA TERRA CO-AXIAL MASTER CHRONOMETER LADIES' 38MM
REF. 220.10.38.20.03.002

Movement: automatic-winding Omega Caliber 8800 Certified Master Chronometer; 55-hour power reserve.
Functions: hours, minutes, seconds.
Case: steel; Ø 38mm; antireflective sapphire crystal; sapphire crystal caseback; water resistant to 15 bar.
Dial: blue.
Bracelet: steel.

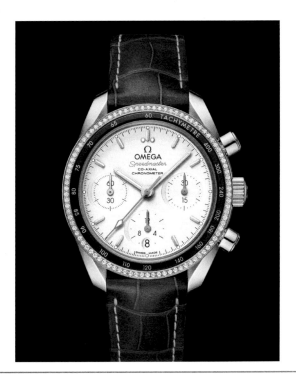

OMEGA

SEAMASTER AQUA TERRA MASTER CHRONOMETER WORLDTIMER 43MM
REF. 220.93.43.22.99.001

Movement: automatic-winding Omega Caliber 8939 Certified Master Chronometer; 60-hour power reserve.
Functions: hours, minutes, seconds; date at 6; worldtimer.
Case: platinum; Ø 43mm; antireflective sapphire crystal; sapphire crystal caseback; screw-down crown; water resistant to 15 bar.
Dial: gray.
Strap: brown leather.

SEAMASTER DIVER 300M CO-AXIAL MASTER CHRONOMETER 42 MM
REF. 210.60.42.20.99.001

Movement: automatic-winding Omega Caliber 8806 Certified Master Chronometer; 55-hour power reserve.
Functions: hours, minutes, seconds.
Case: titanium and Sedna™ gold; Ø 42mm; antireflective sapphire crystal; sapphire crystal caseback; screw-down crown; water resistant to 30 bar.
Dial: gray.
Bracelet: titanium and Sedna™ gold.

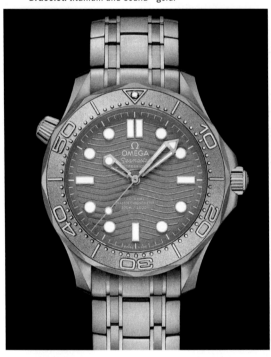

SEAMASTER DIVER 300M CO-AXIAL MASTER CHRONOMETER 42 MM
REF. 210.30.42.20.01.001

Movement: automatic-winding Omega Caliber 8800 Certified Master Chronometer; 55-hour power reserve.
Functions: hours, minutes, seconds; date at 6.
Case: steel; Ø 42mm; antireflective sapphire crystal; sapphire crystal; screw-down crown; water resistant to 30 bar.
Dial: black.
Bracelet: steel.

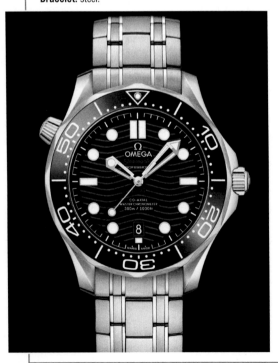

SEAMASTER DIVER 300M CO-AXIAL MASTER CHRONOMETER 42 MM
REF. 210.32.42.20.06.001

Movement: automatic-winding Omega Caliber 8800 Certified Master Chronometer; 55-hour power reserve.
Functions: hours, minutes, seconds; date at 6.
Case: steel; Ø 42mm; antireflective sapphire crystal; sapphire crystal; screw-down crown; water resistant to 30 bar.
Dial: gray.
Strap: blue rubber.

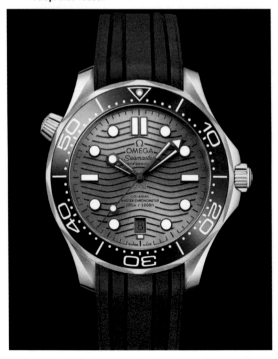

OMEGA

SPEEDMASTER RACING CO-AXIAL MASTER CHRONOMETER CHRONOGRAPH
REF. 329.53.44.51.03.001

Movement: automatic-winding Omega 9901 caliber; 60-hour power reserve; 28,800 vph.
Functions: hours, minutes, seconds; date at 6; chronograph: 12-hour counter at 3, 60-minute counter at 9.
Case: Sedna™ gold; Ø 44.25mm; anti-reflective sapphire crystal; sapphire crystal caseback.
Dial: blue.
Bracelet: blue leather.

SPEEDMASTER DARK SIDE OF THE MOON APOLLO 8 CHRONOGRAPH 44.25MM
REF. 311.92.44.30.01.001

Movement: manually-wound Omega Caliber 1869; 48-hour power reserve.
Functions: hours, minutes, running seconds at 9; chronograph: 30-minute counter at 3, 12-hour counter at 6.
Case: black ceramic.
Dial: black.
Strap: black rally-style leather.
Edition: Dark Side of the Moon Apollo 8.

SPEEDMASTER MOONWATCH PROFESSIONAL CHRONOGRAPH 42MM
REF. 311.30.42.30.01.005

Movement: manually-wound Omega Caliber 1861; 48-hour power reserve.
Functions: hours, minutes, running seconds at 9; chronograph: 30-minute counter at 3, 12-hour counter at 6.
Case: steel; Ø 42mm; hesalite crystal; water resistant to 5 bar.
Dial: black.
Bracelet: steel.

SPEEDMASTER RACING MASTER CHRONOMETER CHRONOGRAPH 44.25MM
REF. 329.30.44.51.04.001

Movement: automatic-winding Omega Caliber 9900 Certified Master Chronometer; 60-hour power reserve.
Functions: hours, minutes; chronograph: 60-minute counter at 9, 12-hour counter at 3; date at 6.
Case: steel; Ø 44.25mm; antireflective sapphire crystal; sapphire crystal caseback; water resistant to 5 bar.
Dial: white.
Bracelet: steel.

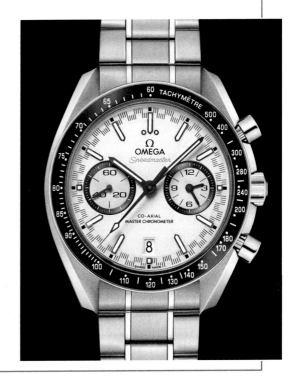

PATEK PHILIPPE
GENEVE

PERENNIAL APPEAL

With a deep well of history to draw upon, Patek Philippe **CELEBRATES ITS IMPECCABLE PAST**, releasing a limited run to honor one of its flagship models, and reimagining a coveted complication for women.

The "divine" proportions behind the famed Golden Ratio led to the timeless look of Patek Philippe's Golden Ellipse, first introduced in 1968. Its 34.50x39.50mm ovoid case melds the classic watch shapes of the circle and the rectangle, bringing a sleek modern vision to an ancient concept of perfection. For the model's 50th anniversary in 2018, the Genevan brand created a limited edition of 100 pieces, embellished by its rare handcrafts specialists. The platinum case of the Golden Ellipse Ref. 5738-50P is echoed in the similarly shaped platinum prong buckle on the shiny black alligator strap. The piece's white-gold dial bears distinctive volutes inspired by Patek Philippe's signature Calatrava cross, a work of art requiring many hours of intense effort from a master engraver. The solid platinum caseback is engraved with "Ellipse d'Or 1968–2018," and two final finishing touches add complementary grace notes: an onyx cabochon set into the crown matches the black enamel dial, and a diamond set into the caseband at 6 o'clock conceptually connects this model to all other platinum Patek Philippe watches. The ultra-thin self-winding caliber 240 within contributes to the case's 6.58mm slenderness, boasting a 22-karat gold minirotor and beating at 21,600 vph. Completing the vision of gentlemanly elegance, each watch is delivered with a matching pair of cufflinks.

◀ **GOLDEN ELLIPSE REF. 5738-50P**
With a minimum power reserve of 48 hours, the self-winding movement at the heart of this Golden Ellipse model reliably powers the minutes and hours display from compact dimensions of Ø 27.5mm and 2.53mm in thickness.

Patek Philippe revisits the Ladies' Chronograph, crafting the model from the ground up with many vintage design references.

Many brands seem to forget that women enjoy chronographs as a complication in their own right. Ten years ago, Patek Philippe brought new attention to the market with a chronograph designed specifically for women. The brand now returns to this inspiration with the revisited Ladies' Chronograph Ref. 7150/250R-001, crafting the model from the ground up, with many references to vintage design. A silvered opaline dial bears Breguet hour and minute hands in a rose gold to match the classic 38mm round, diamond-set case. The model emphasizes its chronograph aspect with round pushbuttons featuring guilloché decoration, a highly legible dial design, and a pulsimeter function that measures the heart rate of its active wearer. The fluted lugs guide the watchcase to the shiny mink gray alligator leather strap with a classic sense of finesse. Inside, the manual-winding CH 29-535 PS caliber powers a column-wheel chronograph with a chronograph hand, 30-minute counter, and subsidiary seconds between 8 and 9 o'clock. The Patek Philippe Seal confirms the ultra-high standards to which this timepiece is held.

The gently chamfered box-style sapphire crystal front and back reveals the intriguing, beautifully decorated inner workings of this landmark model, the only women's chronograph offered by Patek Philippe. A vision of luxury in a life of competence and capability, this Ladies Chronograph is for the woman who knows there is nothing she can't handle.

▶ **LADIES' CHRONOGRAPH REF. 7150/250R-001**
Seventy-two diamonds set around the 18-karat rose-gold dial, as well as 27 on the rose-gold buckle, act as a bright reminder that women like complicated watches, but also appreciate a bit of sparkle.

PATEK PHILIPPE

REF. 5270P-001

Movement: manual-winding CH 29-535 PS Q caliber; Ø 32mm, thickness: 7mm; min. 55-hour power reserve; 456 components; 33 jewels; 28,800 vph; Breguet balance spring.
Functions: hours, minutes; small seconds at 9; perpetual calendar: date by hand at 6, day and month at 12, leap year between 4 and 5; moonphase at 6; day/night indicator between 7 and 8; chronograph: central seconds hand, 30-minute counter at 3.
Case: platinum; Ø 41mm, thickness: 12.4mm; interchangeable full back and sapphire crystal caseback; water resistant to 3atm.

Dial: golden opaline; blackened gold applied numerals.
Strap: hand-stitched shiny chocolate brown alligator leather; fold-over clasp.
Suggested price: $187,110

REF. 5270/1R-001

Movement: manual-winding CH 29-535 PS Q caliber; Ø 32mm, thickness: 7mm; min. 55-hour power reserve; 456 components; 33 jewels; 28,800 vph; Breguet balance spring.
Functions: hours, minutes; small seconds at 9; perpetual calendar: date by hand at 6, day and month at 12, leap year between 4 and 5; moonphase at 6; day/night indicator between 7 and 8; chronograph: central seconds hand, 30-minute counter at 3.
Case: rose gold; Ø 41mm, thickness: 12.4mm; interchangeable full back and sapphire crystal caseback; water resistant to 3atm.

Dial: ebony black sunburst; gold applied hour markers.
Bracelet: rose gold; fold-over clasp.
Suggested price: $192,780

REF. 5531R-001

Movement: automatic-winding R 27 HU caliber; Ø 32mm, thickness: 8.5mm; min. 43-hour power reserve; 462 components; 45 jewels; 21,600 vph; Spiromax® balance spring.
Functions: hours, minutes; world time; day/night indicator for the 24 time zones; minute repeater.
Case: rose gold; Ø 40.2mm, thickness: 11.49mm; interchangeable full back and sapphire crystal caseback; humidity and dust protected.
Dial: cloisonné enamel center; Lavaux landscape; 18K gold dial plate and 24-hour ring; skeletonized lugs.

Strap: hand-stitched shiny chocolate brown alligator leather; fold-over clasp.
Price: available upon request.

REF. 5208R-001

Movement: automatic-winding R CH 27 PS QI caliber; Ø 32mm, thickness: 10.35mm; min. 38-hour power reserve; 719 components; 63 jewels; 21,600 vph; Spiromax® balance spring.
Functions: hours, minutes, small seconds at 6; perpetual calendar: date at 12, day between 10 and 11, month between 1 and 2, leap year at 3; day/night indicator at 9; moonphase at 6; minute repeater; chronograph: central seconds hand, 12-hour counter at 9, 60-minute counter at 3.
Case: rose gold; Ø 42mm, thickness: 15.11mm; interchangeable full back and sapphire crystal caseback; humidity and dust protected.

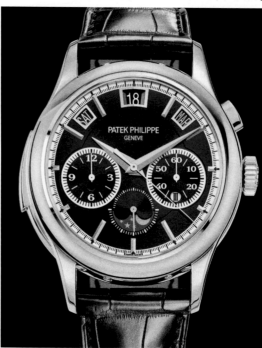

Dial: ebony black sunburst; gold applied hour markers; 18K gold dial plate.
Strap: hand-stitched matte black alligator leather; fold-over clasp.
Price: available upon request.

PATEK PHILIPPE

REF. 5207G

Movement: manual-winding R TO 27 PS QI caliber; thickness: 9.33mm; min. 38-hour power reserve; 549 components; 37 jewels; 21,600 vph; Breguet balance spring.
Functions: hours, minutes; small seconds at 6; perpetual calendar: date at 12, day between 10 and 11, month between 1 and 2, leap year between 4 and 5; moonphase at 6; minute repeater; tourbillon.
Case: white gold; Ø 41mm, thickness: 13.81mm; interchangeable full back and sapphire crystal caseback; humidity and dust protected.
Dial: blue sunburst; gold applied hour markers; 18K gold dial plate.
Strap: hand-stitched shiny navy blue alligator leather; foldover clasp.
Price: available upon request.

REF. 5205G-013

Movement: automatic-winding 324 S QA LU 24H caliber; Ø 32.6mm, thickness: 5.78mm; min. 35-hour power reserve; 356 components; 34 jewels; 28,800 vph; Spiromax® balance spring.
Functions: hours, minutes, seconds; annual calendar: date at 12, day between 10 and 11, month between 1 and 2; moonphase and 24-hour display at 6.
Case: white gold; Ø 40mm, thickness: 11.36mm; sapphire crystal caseback; water resistant to 3atm.
Dial: blue sunburst; gold applied hour markers.
Strap: hand-stitched shiny black alligator leather; prong buckle.
Suggested price: $47,970

REF. 7324R-001

Movement: automatic-winding 324 S C FUS caliber; Ø 31mm, thickness: 4.9mm; min. 35-hour power reserve; 294 components; 29 jewels; 28,800 vph; Spiromax® balance spring.
Functions: hours, minutes, seconds; date by hand at 6; dual time zone; day/night indicator at 4 and 8.
Case: rose gold; Ø 37.5mm, thickness: 10.78mm; sapphire crystal caseback; water resistant to 6atm.
Dial: brown sunburst; gold applied numerals with luminescent coating.
Strap: vintage brown calfskin leather; Clevis prong buckle.
Suggested price: $43,090

REF. 5524R-001

Movement: automatic-winding 324 S C FUS caliber; Ø 31mm, thickness: 4.9mm; min. 35-hour power reserve; 294 components; 29 jewels; 28,800 vph; Spiromax® balance spring.
Functions: hours, minutes, seconds; date by hand at 6; dual time zone; day/night indicator at 4 and 8.
Case: rose gold; Ø 42mm, thickness: 10.78; sapphire crystal caseback; water resistant to 6atm.
Dial: brown sunburst; gold applied numerals with luminescent coating.
Strap: vintage brown calfskin leather; Clevis prong buckle.
Suggested price: $47,630

PATEK PHILIPPE

REF. 5738R-001

Movement: automatic-winding 240 caliber; Ø 27.5mm, thickness: 2.53mm; min. 48-hour power reserve; 161 components; 27 jewels; 21,600 vph; Spiromax® balance spring.
Functions: hours, minutes.
Case: rose gold; 34.5x39.5mm, thickness: 5.9mm; solid caseback; water resistant to 3atm.
Dial: ebony black sunburst; gold applied hour markers; 18K gold dial plate.
Strap: hand-stitched shiny black alligator leather; prong buckle.
Suggested price: $30,850

REF. 5740/1G-001

Movement: automatic-winding 240Q caliber; Ø 27.5mm, thickness: 3.88mm; min. 38-hour power reserve; 275 components; 27 jewels; 21,600 vph; Spiromax® balance spring.
Functions: hours, minutes; perpetual calendar: date at 6, day at 9, month and leap year at 3; moonphase at 6; 24-hour indicator at 9.
Case: white gold; Ø 40mm, thickness: 8.42mm; screw-down crown; sapphire crystal caseback; water resistant to 6atm.
Dial: blue sunburst; gold applied hour markers with luminescent coating.
Bracelet: white gold; Nautilus fold-over clasp.
Suggested price: $119,070

REF. 5968A-001

Movement: automatic-winding CH 28-520 C caliber; Ø 30mm, thickness: 6.63mm; min. 45-hour power reserve; 308 components; 32 jewels; 28,800 vph; Spiromax® balance spring.
Functions: hours, minutes; date at 3; chronograph: central seconds hand, 60-minute counter at 6.
Case: steel; Ø 42.2mm, thickness: 11.9mm; screw-down crown; sapphire crystal caseback; water resistant to 12atm.
Dial: black embossed; gold applied numerals with luminescent coating.
Strap: black composite material; Aquanut fold-over clasp; also comes with an orange composite material strap.
Suggested price: $43,770

REF. 5067A-025

Movement: quartz E23-250 S C caliber; Ø 23.9mm, thickness: 2.5mm; 3-yeary battery life; 80 components; 8 jewels; 32,768 vph.
Functions: hours, minutes, seconds; date at 3.
Case: steel; Ø 35.6mm, thickness: 7.7mm; diamond-set bezel; screw-down crown; solid caseback; water resistant to 12atm.
Dial: embossed blue-gray; gold applied numerals.
Strap: blue-gray composite material; Aquanut fold-over clasp.
Suggested price: $16,220

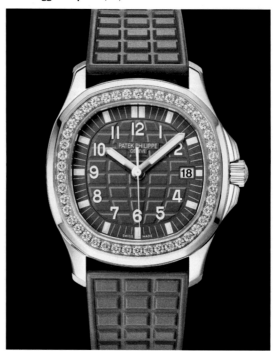

PATEK PHILIPPE

REF. 5327R-01

Movement: automatic-winding 240 Q caliber; Ø 27.5mm, thickness: 3.88mm; min. 38-hour power reserve; 275 components; 27 jewels; 21,600 vph; Spiromax® balance spring.
Functions: hours, minutes; perpetual calendar: date at 6, day at 9, month and leap year at 3; moonphase at 6; 24-hour indicator at 9.
Case: rose gold; Ø 39mm, thickness: 9.71mm; interchangeable full back and sapphire crystal caseback; water resistant to 3atm.
Dial: lacquered ivory; gold applied Breguet numerals.
Strap: hand-stitched shiny dark chestnut alligator leather; fold-over clasp.
Suggested price: $87,320

REF. 5230G-001

Movement: automatic-winding 240 HU caliber; Ø 27.5mm, thickness: 3.88mm; min. 48-hour power reserve; 239 components; 33 jewels; 21,600 vph; Spiromax® balance spring.
Functions: hours, minutes; world time with day/night indicator.
Case: white gold; Ø 38.5mm, thickness: 10.23mm; sapphire crystal caseback; water resistant to 3atm.
Dial: hand-guilloché lacquered charcoal gray; gold applied hour markers.
Strap: hand-stitched shiny black alligator leather; fold-over clasp.
Suggested price: $47,630

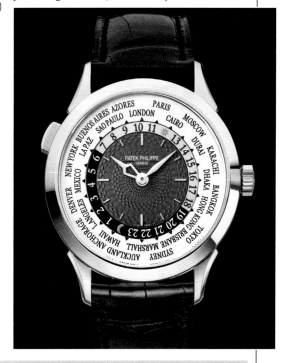

REF. 4910/10A-011

Movement: quartz E15 caliber; 15x13mm, thickness: 1.8mm; 3-year battery life; 57 components; 6 jewels; 32,768 vph.
Functions: hours, minutes.
Case: steel; 25.1x30mm, thickness: 6.8mm; case set with 36 diamonds (~0.45 carat); crown set with an onyx; water resistant to 3atm.
Dial: "timeless" white; diamond hour markers; gold applied Roman numerals.
Bracelet: steel; fold-over clasp.
Suggested price: $12,140

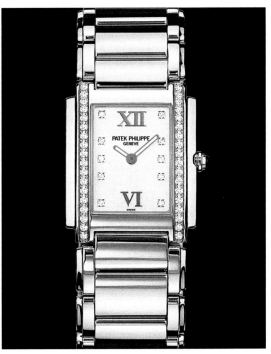

REF. 7200/200R-001

Movement: automatic-winding 240 caliber; Ø 27.5mm, thickness: 2.53mm; min. 48-hour power reserve; 161 components; 27 jewels; 21,600 vph; Spiromax® balance spring.
Functions: hours, minutes.
Case: rose gold; Ø 34.6mm, thickness: 7.37mm; case set with 142 diamonds (~1.08 carats); sapphire crystal caseback; water resistant to 3atm.
Dial: silvery grained; gold applied Breguet numerals.
Strap: hand-stitched shiny royal purple alligator leather; prong buckle set with 26 diamonds (~0.18 carat).
Suggested price: $39,690

PATEK PHILIPPE

REF. 5078G-001

Movement: automatic-winding R 27 PS caliber; thickness: 5.05mm; 39 jewels; 21,600 vph; Patek Philippe seal.
Functions: hours, minutes; small seconds at 6; minute repeater.
Case: white gold; Ø 38mm; interchangeable full caseback and sapphire crystal caseback; humidity- and dust-protected.
Dial: cream enamel; gold applied Breguet numerals.
Strap: hand-stitched chocolate brown alligator leather; fold-over clasp.
Price: available upon request.

REF. 5316P-001

Movement: manual-winding R TO 27 PS QR caliber; thickness: 8.61mm; 28 jewels; 21,600 vph; Patek Philippe seal.
Functions: hours, minutes; small seconds at 6; perpetual calendar: date via central hand, day at 9, month at 3, leap year at 12; moonphase at 6; tourbillon; minute repeater.
Case: platinum; Ø 40.2mm; interchangeable full caseback and sapphire crystal caseback; humidity- and dust-protected.
Dial: black enamel; gold applied hour markers.
Strap: hand-stitched black alligator leather; fold-over clasp.
Price: available upon request.

REF. 5372P-010

Movement: manual-winding CHR 27 525 PS Q caliber; thickness: 7.3mm; min. 38-hour power reserve, max. 48-hour power reserve; 476 components; 31 jewels; 21,600 vph; Patek Philippe seal.
Functions: hours, minutes; small seconds at 3; perpetual calendar: month at 3, leap year between 4 and 5, date via hand at 6, day at 9; day/night indicator between 7 and 8; moonphase at 12; chronograph: 60-minute counter at 9, central split-seconds hands.
Case: platinum; Ø 38.3mm; water resistant to 3atm.
Dial: rose gold; vertical satin-finished with blackened gold applied numerals.
Strap: hand-stitched chocolate brown alligator leather; fold-over clasp.
Price: available upon request.

REF. 5940R-001

Movement: automatic-winding 240 Q caliber; Ø 27.5mm, thickness: 3.88mm; min. 38-hour power reserve, max. 48-hour power reserve; 275 components; 27 jewels; 21,600 vph; Patek Philippe seal.
Functions: hours, minutes; perpetual calendar: date at 6, day at 9, month and leap year at 3; 24-hour indicator at 9; moonphase at 6.
Case: rose gold; 37x44.6mm; interchangeable full caseback and sapphire crystal caseback; water resistant to 3atm.
Dial: silvery grained; gold applied Breguet numerals.
Strap: hand-stitched chocolate brown alligator leather; prong buckle.
Suggested price: $87,320

PATEK PHILIPPE

REF. 5131/1P-001

Movement: automatic-winding 240 HU caliber; Ø 27.5mm, thickness: 3.88mm; 48-hour power reserve; 239 components; 33 jewels; 21,600 vph; Patek Philippe seal.
Functions: hours, minutes; world time; 24-hour day/night indicator for the 24 time zones.
Case: platinum; Ø 39.5mm; sapphire crystal caseback; water resistant to 3atm.
Dial: cloisonné enamel center.
Bracelet: platinum; fold-over clasp.
Suggested price: $130,413

REF. 5170P-001

Movement: manual-winding CH 29 535 PS caliber; Ø 29.6mm, thickness: 5.35mm; 65-hour power reserve; 269 components; 33 jewels; 28,800 vph; Patek Philippe seal.
Functions: hours, minutes; small seconds at 9; chronograph: 30-minute counter at 3, central seconds hand; tachometer scale.
Case: platinum; Ø 39.4mm; sapphire crystal caseback; water resistant to 3atm.
Dial: blue sunburst; baguette diamond hour markers (0.23 carat).
Strap: hand-stitched black alligator leather; fold-over clasp.
Suggested price: $96,392

REF. 5180/1R-001

Movement: automatic-winding 240 SQU/179 caliber; Ø 27.5mm, thickness: 2.53mm; 48-hour power reserve; 159 components; 27 jewels; 21,600 vph.
Functions: hours, minutes.
Case: rose gold; Ø 39mm; sapphire crystal caseback; water resistant to 3atm.
Dial: skeletonized.
Bracelet: rose gold; fold-over clasp.
Suggested price: $98,660

REF. 5396R-015

Movement: automatic-winding 324 S QA LU 24H/303 caliber; Ø 33.3mm, thickness: 5.78mm; min. 35-hour power reserve, max. 45-hour power reserve; 347 components; 34 jewels; 28,800 vph; Patek Philippe seal.
Functions: hours, minutes, seconds; annual calendar: date at 6, day and month at 12; 24-hour display at 6; moonphase at 6.
Case: rose gold; Ø 38.5mm; sapphire crystal caseback; water resistant to 3atm.
Dial: blue sunburst; baguette diamond hour markers (0.26 carat).
Strap: hand-stitched navy blue alligator leather; fold-over clasp.
Suggested price: $53,299

PIAGET

REFINED MODERNITY

Celebrating the elegant Piaget spirit with plenty of sparkle, new models from the Piaget Polo and Limelight collections emphasize two areas of the Maison's expertise: **THE USE OF COLOR AND DIAMOND-SETTING**.

Forty years after its initial launch, the Piaget Polo collection continues to grow and evolve with the times. New 42mm Piaget Polo models, crafted in striking 18-karat pink gold, come in two diamond-set versions. One of them features a simple halo of brilliant-cut diamonds on the emblematic bezel, while the other one is exuberantly adorned with a fully paved dial and bezel. The models come with two alligator straps that add versatility: a sober blue strap sets off the gleam of diamonds, while a deep garnet strap emphasizes the warmth of the pink-gold case. The 1979 Piaget Polo was inspired by the noble sport of the same name.

Powered by the automatic caliber 1110P of Manufacture Piaget, these models express a commitment to beauty inside and out. The movement comprises 280 components and 25 jewels, with haute horology finishings such as circular côtes de Genève, circular-grained mainplate, beveled bridges, sunburst-brushed wheels, blue screws, and an engraved coat of arms on the slate gold oscillating weight. This impeccable piece of machinery beats at 28,800 vph and provides a power reserve of approximately 50 hours.

PIAGET

The new Limelight Gala reveals its scintillating beauty with an entirely new 18-karat white-gold case, a lavish setting of dazzling diamonds, and a white mother-of-pearl dial.

Piaget takes the iconic and dazzling Limelight Gala one step further with the launch of an even more sumptuous 32mm version. Opulent, sinuous, and sensual, the Limelight Gala brings a hint of the outstanding and precious glamour of the era of its birth: the 1970s. The new Limelight Gala reveals its scintillating beauty thanks to an entirely new 18-karat white-gold case, and a lavish gem-setting with dazzling-sized diamonds for a total of 4.8 carats, as well as a white mother-of-pearl dial. As on all Limelight Gala models, the progressive diamond setting enhances the asymmetry of the lugs, which stretch languidly along the strap. This most recent version is adorned with 57 diamonds, each with a unique diameter, from less than 1mm to more than 4mm. Their shimmer is enhanced by the so-called "serti descendu Piaget style," which consists of an openworked setting focusing the spotlight on the precious stones. The engraved arch decoration on the case middle is yet another reference to the sensual curves of Limelight Gala, which create a startling contrast to the black alligator strap with diamond-set ardillon buckle.

▶ **LIMELIGHT GALA**
The iconic Limelight Gala makes a comeback with an even more dazzling version. As with all the Gala watches, the progressive diamond setting enhances the asymmetry of the lugs.

◀ **PIAGET POLO** (*facing page left*)
This Piaget Polo version emphasizes the use of color, featuring a simple halo of 56 brilliant-cut diamonds on the emblematic bezel, with a white dial, pink gold indexes with SuperLumiNova®.

◀ **PIAGET POLO** (*facing page right*)
The Piaget Polo is an utterly contemporary model which distinguishes the collection in the area of gem-setting. This version has a sumptuous, fully paved bezel and dial, set with 496 brilliant-cut diamonds (1.86 carats) and pink-gold indexes.

PIAGET

POSSESSION WATCH – 29 MM   REF. G0A43080

Movement: quartz Piaget Manufacture 56P caliber.
Functions: hours, minutes.
Case: stainless steel; Ø 29mm; set with 1 brilliant-cut diamond (approx. 0.05 ct).
Dial: silvered; set with 11 diamond indexes (approx. 0.07 ct).
Strap: dark blue alligator leather; interchangeable; stainless steel ardillon buckle.
Suggested price: $3,550

POSSESSION WATCH – 29 MM   REF. G0A43085

Movement: quartz Piaget Manufacture Caliber 56P.
Functions: hours, minutes.
Case: 18K white gold; Ø 29mm; set with 162 brilliant-cut diamonds (approx. 1.55 cts).
Dial: silvered; set with 11 diamond indexes (approx. 0.07 ct).
Strap: dark blue alligator leather; interchangeable; stainless steel ardillon buckle.
Suggested price: $20,500

POSSESSION WATCH – 29 MM   REF. G0A43081

Movement: quartz Piaget Manufacture 56P caliber.
Functions: hours, minutes.
Case: 18K pink gold; Ø 29mm; set with 1 brilliant-cut diamond (approx. 0.05 ct).
Dial: silvered; set with 11 diamond indexes (approx. 0.07 ct).
Strap: lapis blue alligator leather; interchangeable; pink-gold-plated ardillon buckle.
Suggested price: $8,050

POSSESSION WATCH – 34 MM   REF. G0A43090

Movement: quartz Piaget Manufacture Caliber 56P.
Functions: hours, minutes.
Case: steel; Ø 34mm; set with 1 brilliant-cut diamond (approx. 0.07 ct).
Dial: silvered; set with 11 diamond indexes (approx. 0.09 ct).
Strap: dark blue alligator leather; interchangeable; stainless steel ardillon buckle.
Suggested price: $4,000

PIAGET

POSSESSION WATCH – 34 MM — REF. G0A43091

Movement: quartz Piaget Manufacture Caliber 56P.
Functions: hours, minutes.
Case: 18K pink gold; Ø 34mm; set with 1 brilliant-cut diamond (approx. 0.07 ct).
Dial: silvered; set with 11 diamond indexes (approx. 0.09 ct).
Strap: lapis blue alligator leather; interchangeable; pink-gold-plated ardillon buckle.
Suggested price: $9,550

POSSESSION WATCH – 34MM — REF. G0A43095

Movement: quartz Piaget Manufacture Caliber 56P.
Functions: hours, minutes.
Case: 18K white gold; Ø 34mm; set with 181 brilliant-cut diamonds (approx. 2.11 cts).
Dial: silvered; set with 11 diamond indexes (approx. 0.09 ct).
Strap: dark blue alligator leather; interchangeable; steel ardillon buckle.
Suggested price: $23,500

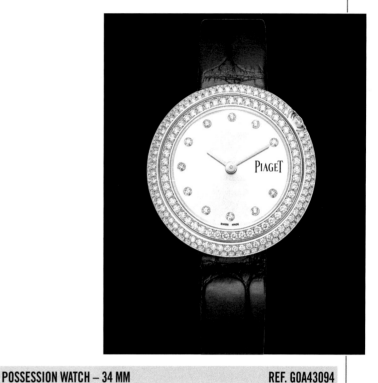

POSSESSION WATCH – 34 MM — REF. G0A43092

Movement: quartz Piaget Manufacture Caliber 56P.
Functions: hours, minutes.
Case: 18K pink gold; Ø 34mm; set with 46 brilliant-cut diamonds (approx. 1.46 cts).
Dial: silvered; set with 11 diamond indexes (approx. 0.09 ct).
Strap: lapis blue alligator leather; interchangeable; pink-gold-plated ardillon buckle.
Suggested price: $16,100

POSSESSION WATCH – 34 MM — REF. G0A43094

Movement: quartz Piaget Manufacture Caliber 56P.
Functions: hours, minutes.
Case: 18K white gold; Ø 34mm; set with 46 brilliant-cut diamonds (approx. 1.46 cts).
Dial: silvered; set with 11 diamond indexes (approx. 0.09 ct).
Strap: dark blue alligator leather.
Suggested price: $17,000

PIAGET

LIMELIGHT STELLA – 36 MM — REF. G0A40039

Movement: automatic-winding 580P caliber; Ø 26mm, thickness: 4.51mm; 42-hour power reserve; 21,600 vph.
Functions: hours, minutes; moonphase at 12.
Case: 18K white gold; Ø 36mm, thickness: 10.3mm; set with diamonds.
Dial: white mother-of-pearl; baguette-cut diamonds in center.
Strap: gray alligator leather; gray ardillon buckle.
Suggested price: $248,000

LIMELIGHT STELLA – 36 MM — REF. G0A40111

Movement: automatic-winding 584P caliber; 42-hour power reserve; black oscillating weight.
Functions: hours, minutes, seconds; moonphase at 12.
Case: 18K white gold; Ø 36mm; set with 126 brilliant-cut diamonds (approx. 0.65 carat); sapphire crystal caseback.
Dial: white; white-gold hour markers; moonphase underlined with 14 diamonds (approx. 0.06 carat).
Strap: blue alligator leather.
Suggested price: $30,700

LIMELIGHT STELLA – 36 MM — REF. G0A40123

Movement: automatic-winding 584P caliber; 42-hour power reserve; pink-gold-colored oscillating weight.
Functions: hours, minutes, seconds; moonphase at 12.
Case: 18K pink gold; Ø 36mm; set with 126 brilliant-cut diamonds (~0.65 carat); sapphire crystal caseback.
Dial: white; pink-gold hour markers; moonphase underlined with 14 diamonds (~0.06 carat).
Strap: taupe alligator leather.
Suggested price: $29,600

LIMELIGHT STELLA – 36 MM — REF. G0A40110

Movement: automatic-winding 584P caliber; 42-hour power reserve; pink-gold-colored oscillating weight.
Functions: hours, minutes, seconds; moonphase at 12.
Case: 18K pink gold; Ø 36mm; sapphire crystal caseback.
Dial: white; pink-gold hour markers; moonphase underlined with 14 diamonds (~0.06 carat).
Strap: brown alligator leather.
Suggested price: $21,200

PIAGET

PIAGET LIMELIGHT GALA – 26 MM — REF. G0A42213

Movement: quartz 59P Piaget Caliber.
Functions: hours, minutes.
Case: 18K pink gold; Ø 26mm; set with 60 brilliant-cut diamonds (approx. 0.92 ct).
Dial: silver; pink-gold-colored Roman numerals.
Bracelet: 18K pink-gold Milanese mesh.
Suggested price: $25,400

PIAGET LIMELIGHT GALA – 26 MM — REF. G0A42150

Movement: quartz Piaget 59P Caliber.
Functions: hours, minutes.
Case: 18K white gold; Ø 26mm; set with 60 brilliant-cut diamonds (approx. 0.92 ct).
Dial: silver.
Strap: black satin; clasp set with 1 brilliant-cut diamond (approx. 0.01 ct).
Suggested price: $21,700

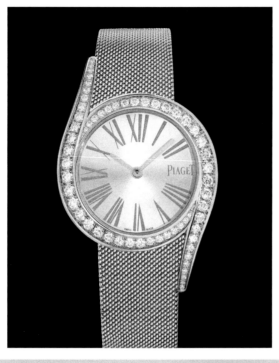

LIMELIGHT GALA MILANESE – 32 MM — REF. G0A41213

Movement: quartz 690P caliber.
Functions: hours, minutes.
Case: 18K pink gold; Ø 32mm; set with 62 brilliant-cut diamonds (approx. 1.75 cts).
Dial: mother-of-pearl; pink-gold-colored Roman numerals.
Bracelet: 18K pink-gold Milanese mesh; integrated sliding clasp engraved with the Piaget "P".
Suggested price: $37,200

LIMELIGHT GALA MILANESE – 32 MM — REF. G0A41212

Movement: quartz 690P caliber.
Functions: hours, minutes.
Case: 18K white gold; Ø 32mm; set with 62 brilliant-cut diamonds (approx. 1.75 cts).
Dial: mother-of-pearl; silver-toned Roman numerals.
Bracelet: 18K white-gold Milanese mesh; integrated sliding clasp.
Suggested price: $38,700

PIAGET

PIAGET ALTIPLANO ULTIMATE 910P   REF. G0A43120

Movement: automatic-winding Piaget Manufacture Caliber 910P; merger between the caliber and exterior elements; 50-hour power reserve; 238 components (case + movement); 30 jewels; 21,600 vph; finishing operations: circular satin-brushed caseback, sunburst satin-brushed bridges, chamfered bridges, sunburst or circular satin-finished wheels, black-coated screws, dedicated index-assembly with "P" as the Piaget signature.
Functions: hours, minutes.
Case: 18K pink gold; Ø 41mm, thickness: 4.30mm; circular satin-brushed caseback.
Strap: black alligator leather; 19K pink-gold ardillon buckle.
Note: world's thinnest mechanical automatic watch.
Suggested price: $27,300

PIAGET ALTIPLANO ULTIMATE 910P   REF. G0A43121

Movement: automatic-winding Piaget Manufacture Caliber 910P; merger between the caliber and exterior elements; 50-hour power reserve; 238 components (case + movement); 30 jewels; 21,600 vph; finishing operations: circular satin-brushed caseback, sunburst satin-brushed bridges, chamfered bridges, sunburst or circular satin-finished wheels, black-coated screws, dedicated index-assembly with "P" as the Piaget signature.
Functions: hours, minutes.
Case: 18K white gold; Ø 41mm, thickness: 4.30mm; circular satin-brushed caseback.
Strap: black alligator leather; 19K white-gold ardillon buckle.
Note: world's thinnest mechanical automatic watch.
Suggested price: $28,400

PIAGET ALTIPLANO DATE   REF. G0A38131

Movement: automatic-winding 1205P caliber; Ø 29.9mm, thickness: 3mm; 44-hour power reserve; 27 jewels; 21,600 vph; finishing: circular Côtes de Genève, circular-grained mainplate, beveled bridges, sunburst-brushed wheels, blued screws; 22K pink-gold oscillating weight.
Functions: hours, minutes; small seconds at 5; date at 9.
Case: 18K pink gold; Ø 40mm; sapphire crystal caseback.
Dial: silvered; black baton-shaped hour markers.
Strap: brown alligator leather; 18K pink-gold ardillon buckle.
Suggested price: $23,800

PIAGET ALTIPLANO DATE   REF. G0A38130

Movement: automatic-winding 1205P caliber; Ø 29.9mm, thickness: 3mm; 44-hour power reserve; 27 jewels; 21,600 vph; finishing: circular Côtes de Genève, circular-grained mainplate, beveled bridges, sunburst-brushed wheels, blued screws; 22K pink-gold oscillating weight.
Functions: hours, minutes; small seconds at 5; date at 9.
Case: 18K white gold; Ø 40mm; sapphire crystal caseback.
Dial: silvered; black baton-shaped hour markers.
Strap: black alligator leather; 18K white-gold ardillon buckle.
Suggested price: $24,700

PIAGET

PIAGET ALTIPLANO HIGH JEWELLERY — REF. G0A41122

Movement: manual-winding 900D caliber.
Functions: hours, minutes.
Case: 18K white gold; Ø 38mm; set with diamonds.
Dial: pavé-set with diamonds.
Strap: black alligator leather.
Note: set with 656 brilliant-cut diamonds (approx. 3.06 cts), 1 diamond (approx. 0.03 ct) and 76 baguette-cut diamonds (approx. 2.73 cts).
Suggested price: $192,000

PIAGET ALTIPLANO – 38 MM — REF. G0A40013

Movement: manual-winding 900P caliber.
Functions: hours, minutes.
Case: 18K pink gold; Ø 38mm, thickness: 3.65mm; set with 78 brilliant-cut diamonds (0.71 carat).
Dial: silvered.
Strap: black alligator leather.
Note: the thinnest hand-wound mechanical watch in the world.
Suggested price: $31,300

PIAGET ALTIPLANO AUTOMATIC GEM-SET SKELETON – 40 MM REF. G0A38125

Movement: automatic-winding 1200D caliber; Ø 31.9mm; 44-hour power reserve; 26 jewels; 21,600 vph; finishing: 14K gold mainplate and bridges gem-set on the upper part, sunray satin-brushed bridges, beveled and hand-drawn mainplate and bridges with sunburst-brushed wheels, black-coated platinum micro-rotor engraved with the Piaget coat-of-arms.
Functions: hours, minutes.
Case: 18K white gold; Ø 40mm; bezel set with 40 baguette-cut diamonds (~3.2 carats); case, crown, lugs and sapphire crystal caseback set with 347 brilliant-cut diamonds (~1.4 carats).
Dial: skeletonized; set with 259 brilliant-cut diamonds (~0.8 carat) and 11 black sapphire cabochons (~0.1 carat).
Strap: black alligator leather; 18K white-gold triple-folding clasp set with 24 brilliant-cut diamonds (~0.06 carat).
Suggested price: $192,000

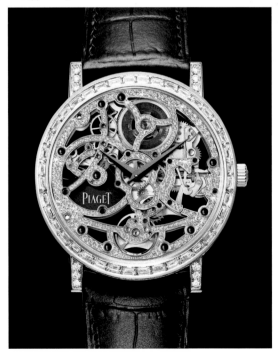

PIAGET ALTIPLANO 900P — REF. G0A39112

Movement: manual-winding 900P caliber; 48-hour power reserve; 145 components; 20 jewels; 21,600 vph; finishing: circular, satin-brushed caseback, sunburst-brushed bridges, beveled bridges, sunburst or circular satin-brushed wheels, slate gray-colored screws.
Functions: hours, minutes.
Case: 18K white gold; Ø 38mm, thickness: 3.65mm; set with 78 brilliant-cut diamonds (~0.71 carat).
Dial: silvered.
Strap: black alligator leather; 18K white-gold pin buckle.
Note: the thinnest hand-wound mechanical watch in the world.
Suggested price: $32,300

PIAGET

PIAGET POLO S – 42 MM — REF. G0A43001

Movement: automatic-winding Piaget Manufacture Caliber 1110P.
Functions: hours, minutes, seconds; date at 6.
Case: steel; Ø 42mm; sapphire crystal; sapphire crystal caseback; water resistant to 100m.
Dial: blue; silvered appliqué indexes with SuperLumiNova.
Strap: blue alligator leather; steel folding clasp.
Suggested price: $8,100

PIAGET POLO S – 42 MM — REF. G0A41003

Movement: automatic-winding 1110P caliber; slate gray oscillating weight.
Functions: hours, minutes, seconds; date at 6.
Case: steel; Ø 42mm, thickness: 9.4mm; sapphire crystal caseback; water resistant to 10atm.
Dial: slate gray; silvered appliqué indexes with SuperLumiNova.
Bracelet: steel; integrated folding clasp.
Suggested price: $9,900

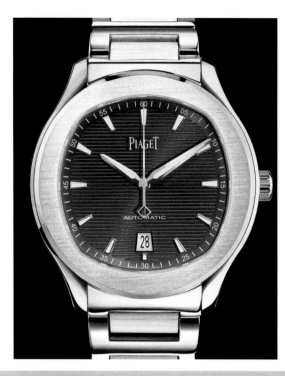

PIAGET POLO S – 42 MM — REF. G0A43002

Movement: automatic-winding Piaget Manufacture Caliber 1160P.
Functions: hours, minutes; chronograph: 30 minute counter at 3, 12 hour counter at 9; date at 6.
Case: steel; Ø 42mm; sapphire crystal; sapphire crystal caseback; water resistant to 100m.
Dial: blue; silvered appliqué indexes with SuperLumiNova.
Strap: blue alligator leather; steel folding clasp.
Suggested price: $11,200

PIAGET POLO S – 42 MM — REF. G0A42005

Movement: automatic-winding 1160P caliber; slate gray oscillating weight.
Functions: hours, minutes; date at 6; chronograph: 12-hour counter at 9, 30-minute counter at 3, central seconds hand.
Case: steel; Ø 42mm, thickness: 11.2mm; sapphire crystal caseback; water resistant to 10atm.
Dial: slate gray; silvered appliqué indexes with SuperLumiNova.
Bracelet: steel; integrated folding clasp.
Suggested price: $13,200

PIAGET

PIAGET EMPERADOR COUSSIN TOURBILLON SKELETON — REF. G0A40042

Movement: automatic-winding Piaget Manufacture Caliber 1270S; Ø 34.98mm, thickness: 5.05mm; 40-hour power reserve; 35 jewels; 21,600 vph.
Functions: hours, minutes; tourbillon at 1.
Case: 18K red gold; Ø 46.5mm, thickness: 8.85mm; sapphire crystal; sapphire crystal caseback.
Strap: black or brown alligator leather; 18K pink-gold folding buckle.
Suggested price: $306,000
Also available: 18K white gold.

BLACK TIE PIAGET EMPERADOR COUSSIN XL — REF. G0A38058

Movement: manual-winding 1270P; Ø 34.9mm, thickness: 5.5mm; 42-hour power reserve; 21,600 vph.
Functions: hours, minutes; power reserve indicator.
Case: Ø 46.5mm, thickness: 10.4mm; 18K red gold; diamond-pavé; water resistant to 3atm.
Dial: skeletonized.
Strap: brown alligator leather; deployant buckle.
Note: limited edition piece.
Suggested price: $232,000

PIAGET EMPERADOR COUSSIN MINUTE REPEATER — REF. G0A38019

Movement: automatic-winding 1290P caliber; thickness: 4.8mm; 40-hour power reserve; 44 jewels; 21,600 vph.
Functions: hours, minutes; minute repeater.
Case: 18K pink gold; Ø 48mm, thickness: 9.4mm; sapphire crystal; gilded decorative fillet with the inscription "Toujours faire mieux que nécessaire"; sapphire crystal caseback.
Dial: skeletonized; gilded hour markers.
Strap: brown alligator leather; 18K pink-gold triple-folding clasp.
Note: world's thinnest self-winding minute repeater watch.
Suggested price: $290,000

PIAGET EMPERADOR COUSSIN TOURBILLON SKELETON — REF. G0A40041

Movement: automatic-winding 1270S caliber; Ø 34.98mm, thickness: 5.05mm; 40-hour power reserve; 225 components; 35 jewels; 21,600 vph; finishing: haute horology finishing, hand-beveled and polished satin-brushed mainplate, hand-beveled and polished satin-brushed bridges, hand-drawn mainplate and bridges, black PVD screws, satin-brushed wheels, beveled and satin-brushed barrel, black PVD-coated, satin-brushed, hand-beveled and polished oscillating weight.
Functions: hours, minutes; tourbillon at 1.
Case: 18K white gold; Ø 46.5mm, thickness: 8.85mm; sapphire crystal caseback.
Dial: skeletonized.
Strap: black alligator leather; 18K white-gold double-folding clasp.
Suggested price: $226,000

RICHARD MILLE

CHALLENGING FORM AND FUNCTION

To everyone attracted to the refinements of fine watchmaking, the experience of meeting a person is very similar to that of taking in a horological creation. The shape of the watch case can tell us a story in mere seconds about the movement held within its exterior, its complexity and features, the type of watch it is, the group of people for whom it was destined, and much more. In fact, the barely visible part of the watch case peeking out from under a cuff or sleeve along with the strap attached to it starts to tell us a story even before the entire case comes into view. It is therefore no wonder that at Richard Mille, a tremendous amount of attention is given to every aspect and detail of an individual model's particular shape and volume—from the design to the actual milling and execution of each case delivered. It is also the reason behind the fact that the cases of Richard Mille, from the first RM 001 to all that followed, have been widely recognized as the most complex and costly to produce within Switzerland—and have remained so to this very day.

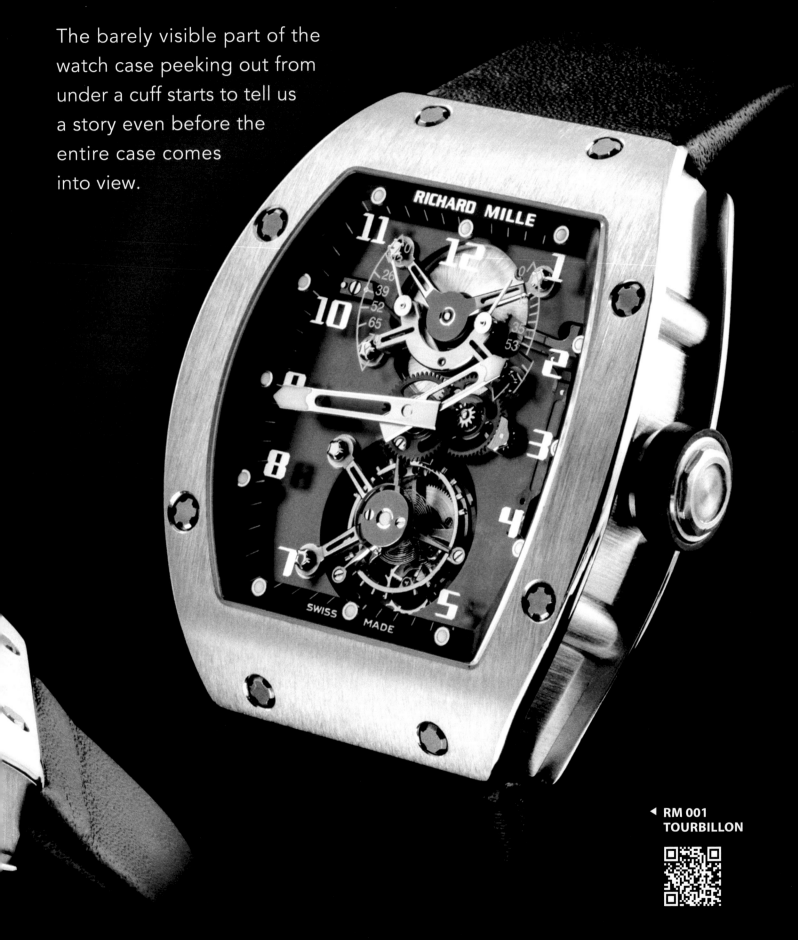

RICHARD MILLE

The barely visible part of the watch case peeking out from under a cuff starts to tell us a story even before the entire case comes into view.

◀ RM 001
TOURBILLON

▶ JACKIE CHAN AND THE RM 025 TOURBILLON CHRONOGRAPH DIVER'S WATCH

RICHARD MILLE

◀ RM 025
TOURBILLON
CHRONOGRAPH
DIVER'S WATCH

BIRTH OF A 21ST CENTURY ICON

It was decades of experience that resulted in that first iconic tonneau-inspired Richard Mille case design that would be followed by many variations and new case forms as time passed. Its realization was a true hands-on development. As Mille explained, "For me, a case is never a simple housing for a watch movement; it is nothing less than a piece of sculpture, despite its diminutive form. And exactly like sculpture, it begins with the definition of volume, the demarcation of negative and positive space, as well as the realization of where these areas intersect one another. And this innately implies that a great design will necessarily contain a certain amount of visual tension in order to realize a shape that will remain attractive and fascinating even after years of long-term viewing." In a moment of high inspiration late at night, Richard Mille began playfully carving a bar of soap into a wrist-hugging form without any flat areas or sharp corners. This first al fresco study soon gave way to a clearer definition of the curvature and volumetric definition of the watch using sheets of plastified paper, cut to shape and glued. Mille continued, "Once the overall external spatial definitions were expressed in the case's form, as well as the ergonomic questions of how the case would interact with the wrist, I had to really examine the details of the interior volume in connection with the movement's requirements, as well as the visual aspects of the same. Since I wanted to use this volume to create depth for exposing the movement in a dramatic manner, there were many demands that needed to be fulfilled."

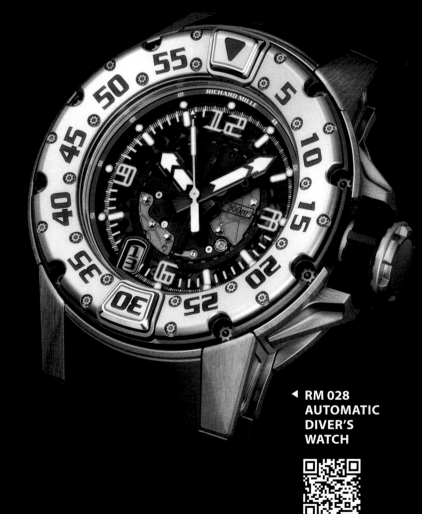

◀ RM 028
AUTOMATIC
DIVER'S
WATCH

RICHARD MILLE

ONE ICON IS NEVER ENOUGH

After the brilliant success of that first case design and all its various iterations—including a ladies' model—the sea introduced a new creative challenge. Why? Because watches destined for diving cannot be made highly water resistant if the watch glass has "corners" of any kind. This meant that the challenge for Richard Mille was the creation of a round watch without succumbing to the classical and existing solutions developed over the course of a hundred years of history. The result was the RM 025, a rare tourbillon chronograph model and the first round wristwatch to grace the collection. It was quickly followed by an automatic diver's wristwatch, the RM 028, and later other round models such as the RM 031, RM 032, RM 033, RM 60-01 and many others, firmly entrenching the round case form into the brand's various collections.

◀ RM 031
HIGH PERFORMANCE

▶ RM 033
AUTOMATIC
EXTRA FLAT

The key to the success of these round Richard Mille interpretations was the use of a "deconstructionist" approach to the case design. The rhythmic definition along the sides of these round cases, the separation of the lugs from the case band—visually and structurally—was a radical departure from the classical usage of other brands. A new type of fixed yet rotatable bezel construction, created through the application of specialized torque screws and attachment methods developed by the brand, makes these round cases absolutely unique in their class.

◀ RM 60-01
AUTOMATIC CHRONOGRAPH
REGATTA

RICHARD MILLE

◀ SKIPPER PIERRE CASIRAGHI AND HIS RM 60-01 AUTOMATIC CHRONOGRAPH REGATTA

RICHARD MILLE

SQUARING THE CIRCLE

After conquering the circle, the next design challenge was the re-interpretation of another classical watch shape: the four-sided square or rectangle. The result was the RM 016, an automatic watch with an ultra-thin profile exuding a 1970s retro mood along with its wrist-hugging curves and fashionable wide strap available in many colors and materials. Mille elaborated, "Rectangular or square-shaped watches have always been produced since the wristwatch existed, and their popularity has generally followed various waves of fashion over the course of almost 150 years. My challenge was to make it contemporary through a total reconception. And of course, with my no-compromise approach, this new shape also had to have the same ergonomic, costly-to-produce curving case shape as found in all the brand's watches."

▲ JAMAICAN SPRINTER YOHAN BLAKE WEARING THE RM 017 TOURBILLON EXTRA FLAT

RICHARD MILLE

The RM 020 Tourbillon Pocket Watch, a homage to the lost art of that definitive gentleman's accoutrement, followed the RM 016 later with a larger interpretation of this rectangular format to bring the pocket-watch firmly into the 21st century. With two winding barrels and a 10-day power reserve tourbillon movement built on a carbon nanofiber baseplate, together with a complex chain and attachment system specially developed for it, the RM 022 is unique, not only within the Richard Mille collections, but also within the entire field of contemporary Swiss watchmaking available today.

Cementing the rectangular shape further into the heart of the brand's watch case design palette and visual language, the RM 017 took on the remarkable challenge of creating a tourbillon within the same case design of the abovementioned RM 016, with only an additional 0.5mm added to the total case height.

◀ RM 020
TOURBILLON
POCKET WATCH

THE ONGOING SEARCH FOR PERFECTION

Total perfection is unachievable, requiring a continuous striving and devotion to the highest standards. This means that Richard Mille approaches all the details of case-making with a view to utilizing the newest technological developments in the machining and manufacturing processes for watch cases, as well as their rigorous quality control. The same devotion to creative details can also be found in the newest iterations even of existing, iconic designs, as the search for excellence is endless. In this vein, the RM 27-02 Tourbillon Rafael Nadal introduced a stepped bezel—by introducing a material called Quartz TPT©—to the iconic tonneau case shape, accentuating the sportive character of this ultra-light timepiece. It is a characteristic example of the brand's dedication to constant improvement and innovation, always providing clients with fresh design elements in an uncompromising timepiece.

◀ RM 27-02
TOURBILLON RAFAEL NADAL

ROGER DUBUIS
DARE TO BE RARE

WITNESS TO MADNESS

EVERY DAY IS A COMPETITION DAY for Roger Dubuis, filled with challenges to be taken on and ambitious goals to be reached. Like Lamborghini Squadra Corse, the Maison is firmly committed to cutting-edge performance, ferocious groundbreaking technology and super-sleek aesthetics, all fuelled by a high-impact R&D vision and focused on delivering bold customer experiences. This set of common denominators put them on a fast track to a roaring encounter where sparks were bound to fly.

For individuals committed to pushing their own limits and defying all known rules, Lamborghini's Squadra Corse motorsport department and Manufacture Roger Dubuis have opted to pool their knowledge and expertise. When visionary engineers meet with incredible watchmakers, the inevitable surge of adrenaline results in models "powered by raging mechanics," the first of which was unleashed on September 20, 2017. More than a year into these high-octane joint endeavors, the Italian aesthetic flair continues to merge with Swiss precision and mechanical instruments.

In 2018, inspired by Lamborghini Squadra Corse and its unique automobile aesthetic showcased in the Huracán, the Manufacture's watchmakers developed a timepiece brimming with racing design codes. One of the most striking features of the aggressive Excalibur Huracán timepiece is its brand-new "engine": the RD630 movement with its 12° inclined balance wheel representing the caliber signature associated with Lamborghini. Taking strong visual cues from the shapes on which the supercar is based, this timepiece notably gives pride of place to the hexagon. A versatile geometrical figure used to construct the volume of the supercar, it appears throughout the timepiece in various guises and is notably reproduced to symbolize strut bars. The stand-out half hexagon appearing on the louvered air intakes of the Lamborghini Huracán are repeated as miniature versions visible through the openworked dial of the Excalibur Huracán timepiece. Other special decoration includes a new crown inspired by racing nuts on supercar wheels, a twin-barrel "energy tank" and a multi-material "spoiler," used for the decorative openworked bridges. Viewed from the back, even the complete circular weight of the automatic movement emulates the design of the wheel rims of Lamborghini's Huracán family. Offering the same blend of performance, lightness and security as its namesake supercar, the Excalibur Huracán timepiece sports distinctive titanium DLC livery complete with signature bright blue accents.

ROGER DUBUIS

When the legendary supplier of racing tires meets the iconic and uncompromising Roger Dubuis watchmaker, the result is a stunning performance that creates the uniqueness of Excalibur Pirelli collection.

The Swiss watchmaker has also entered into an exclusive partnership with Pirelli: world pioneer in technical and industrial innovations and sole supplier of premium tires for the world's flagship motorsport competitions. Playing with the famous color codes of Pirelli, the surprise comes from the strap featuring rubber inlays from certified Pirelli winning tires having competed in real races—and adorned with legendary tread motifs reproducing the profile of a Pirelli Cinturato™ intermediate tire. The robust 45mm skeleton case with its black DLC titanium elements and contrasting with Pirelli mythical tire colors provides a perfect frame for the Calibre RD820SQ. The design and height of the famous Roger Dubuis microrotor effectively enhance the spectacular 3D effect favored by the Manufacture, as does the raised position of the Astral Skeleton. As well as enjoying the key "Pirelli" signature characteristics linking their high-octane timepieces with a landmark motorsports victory, the production of this timepiece is limited to 88 units.

▲ **EXCALIBUR SPIDER PIRELLI**
Excalibur Spider is the only timepiece on the market to be 100% skeletonized, from the case to the movement. It is Roger Dubuis's emblematic collection with an explicitly stated sport aesthetic and the famous Astral skeleton signature.

▶ **AUTOMATIC RD820SQ SERIES**
The automatic movement is highlighted by the astral skeleton structure and the micro-rotor, offering a spectacular view on the mechanism. Not only aesthetically pleasing, the micro-rotor offers a winding ratio as good as the best unidirectional large oscillating weight..

◀ **EXCALIBUR HURACAN**
The second caliber developed for the partnership with Lamborghini, this new engine offers a 12° angle balance escapement with an automatic-winding mechanism displaying a rim-like rotor. The upper caliber features a strut-bar designed bridge recalling the ones of the V10 engine of the Lamborghini Huracan supercar.

ROGER DUBUIS

EXCALIBUR HURACAN — REF. EX0749

Movement: automatic-winding Caliber RD630 Manufacture Roger Dubuis; 233 components; 28,800 vph; 60-hour power reserve.
Functions: hours, minutes, seconds; date.
Case: titanium with black DLC coating overmolded with blue rubber; Ø 45mm; fluted titanium bezel with blue lacquered markings; black DLC titanium crown with blue rubber ring; antireflective sapphire crystal with blue metallization; black DLC titanium caseback with sapphire crystal; water resistant to 5 bar.
Dial: skeletonized; black lower flange and rhodium-plated upper flange; rhodium-plated indexes filled with white SuperLumiNova; 18K black PVD coated gold hands with blue transfers and white SuperLumiNova tips; aluminum blue varnished seconds hand.
Strap: black rubber base with blue Alcantara® inlay; blue stitching; quick release system.
Suggested price: $47,200

EXCALIBUR HURACAN — REF. EX0748

Movement: automatic-winding Caliber RD630 Manufacture Roger Dubuis; 233 components; 28,800 vph; 60-hour power reserve.
Functions: hours, minutes, seconds; date.
Case: titanium with black DLC coating overmolded with black rubber; Ø 45mm; fluted titanium bezel with black lacquered markings; titanium crown with black rubber ring; antireflective sapphire crystal; titanium caseback with sapphire crystal; water resistant to 5 bar.
Dial: skeletonized with black lower flange and rhodium-plated upper flange; rhodium-plated indexes filled with white SuperLumiNova; 18K black PVD coated gold hands with black transfers and white SuperLumiNova tips; aluminum gray varnished second hand.
Strap: black rubber base and gray Alcantara® inlay; black stitching; quick release system.
Suggested price: $47,200

EXCALIBUR HURACAN — REF. EX0750

Movement: automatic-winding Caliber RD630 Manufacture Roger Dubuis; 233 components; 28,800 vph; 60-hour power reserve.
Functions: hours, minutes, seconds; date.
Case: 18K pink gold; Ø 45mm; with titanium container overmolded with black rubber; fluted titanium bezel with black lacquered markings; titanium crown with black rubber ring; antireflective sapphire crystal; titanium caseback with sapphire crystal; water resistant to 5 bar.
Dial: skeletonized; pink-gold-plated lower flange and black upper flange; pink-gold-plated indexes filled with white SuperLumiNova; 18K black PVD coated gold hands with pink-gold transfers and white SuperLumiNova tips; aluminum gold-plated second hand.
Strap: black rubber base and gray Alcantara® inlay; black stitching; quick release system.
Suggested price: $68,500

EXCALIBUR AVENTADOR S — REF. EX0686

Movement: manual-winding Caliber RD103SQ Manufacture Roger Dubuis; 313 components; 57,600 vph; 40-hour power reserve.
Functions: hours, minutes; jumping seconds; power reserve indicator.
Case: C-SMC carbon; Ø 45mm; titanium container overmolded with blue rubber; fluted C-SMC carbon bezel with blue markings; titanium black DLC crown overmolded with blue rubber; antireflective sapphire crystal with red metalization; black DLC titanium caseback with sapphire crystal; water resistant to 5 bar.
Dial: skeletonized with C-SMC carbon; black and red flange; rhodium-plated indexes filled with white SuperLumiNova; 18K black PVD coated gold hands with red transfer and white SuperLumiNova tips.
Strap: black rubber base and neptune blue inlay; red stitching; RD pattern on the inside of the strap; quick release system.
Suggested price: $210,000

ROGER DUBUIS

EXCALIBUR PIRELLI – AUTOMATIC SKELETON REF. EX0745

Movement: automatic-winding Caliber RD820SQ Manufacture Roger Dubuis; 166 components; 28,800 vph; 60-hour power reserve.
Functions: hours, minutes.
Case: titanium; Ø 45mm ; black DLC titanium container overmolded with white rubber; fluted titanium black DLC bezel with white lacquered lines; black DLC titanium crown overmolded with white rubber; antireflective sapphire crystal; titanium black DLC caseback with sapphire crystal; water resistant to 5 bar.
Dial: black and white lower flange; black upper flange with rhodium-plated indexes filled with white SuperLumiNova; 18K black PVD coated gold hands with black transfer and white SuperLumiNova tips.
Strap: black rubber base; black rubber Pirelli-winning tire inlay; white stitching; quick release system.
Suggested price: $73,500

EXCALIBUR PIRELLI – AUTOMATIC SKELETON REF. EX0746

Movement: automatic-winding Caliber RD820SQ Manufacture Roger Dubuis; 166 components; 28,800 vph; 60-hour power reserve.
Functions: hours, minutes.
Case: titanium; Ø 45mm ; black DLC titanium container overmolded with blue rubber; fluted titanium black DLC bezel with blue lacquered lines; black DLC titanium crown overmolded with blue rubber; antireflective sapphire crystal; titanium black DLC caseback with sapphire crystal; water resistant to 5 bar.
Dial: black and blue lower flange; black upper flange with rhodium-plated indexes filled with white SuperLumiNova; 18K black PVD coated gold hands with blue transfer and white SuperLumiNova tips.
Strap: black rubber base; black rubber Pirelli-winning tire inlay; blue stitching; quick release system.
Suggested price: $73,500

EXCALIBUR 45 – DOUBLE TOURBILLON REF. EX0395

Movement: manual-winding Caliber RD01SQ Skeleton Double Flying Tourbillon; 301 components; 43,200 vph; 50-hour power reserve.
Functions: hours, minutes; double flying tourbillon at 5 and 7.
Case: 18K pink gold; Ø 45mm; Excalibur signature fluted bezel; water resistant to 5 bar.
Dial: skeletonized with charcoal gray flange; white minute-circle and Roger Dubuis transfers; pink-gold indexes.
Strap: dark brown genuine alligator leather.
Suggested price: $311,000

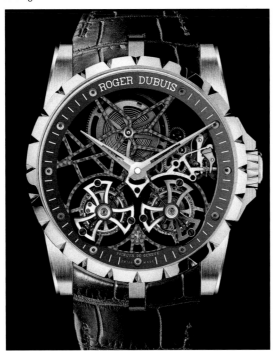

EXCALIBUR 42 – TOURBILLON REF. EX0392

Movement: manual-winding flying tourbillon Caliber RD505SQ; 179 components; 21,600 vph; 60-hour power reserve.
Functions: hours, minutes; flying tourbillon at 7.
Case: 18K pink gold; Ø 42mm; Excalibur signature fluted bezel; water resistant to 3 bar.
Dial: black flange with white minute-circle; Roger Dubuis transfers; pink gold indexes.
Strap: brown alligator leather with brown stitching.
Suggested price: $162,000

ROGER DUBUIS

EXCALIBUR 42 – AUTOMATIC SKELETON   REF. EX0473

Movement: automatic-winding Caliber RD820SQ Manufacture Roger Dubuis; 166 components; 28,800 vph; 60-hour power reserve.
Functions: hours, minutes.
Case: black DLC-coated titanium; Ø 42mm; Excalibur signature fluted bezel; black DLC-coated titanium crown; antireflective sapphire crystal; water resistant to 3 bar.
Dial: skeletonized; charcoal gray flange; white minute-circle and Roger Dubuis transfers; white gold hour markers.
Strap: charcoal black genuine alligator leather.

Suggested price: $63,500

EXCALIBUR 42 – AUTOMATIC SKELETON   REF. EX0777

Movement: automatic-winding Caliber RD820SQ Manufacture Roger Dubuis; 166 components; 28,800 vph; 60-hour power reserve.
Functions: hours, minutes.
Case: carbon; Ø 42mm; Excalibur signature fluted carbon bezel; carbon crown; antireflective sapphire crystal; water resistant to 3 bar.
Dial: skeletonized; charcoal gray flange; red minute-circle and Roger Dubuis transfers; white-gold hour markers.
Strap: black rubber; red stitching.

Suggested price: $67,500

EXCALIBUR 42 – AUTOMATIC SKELETON   REF. EX0793

Movement: automatic-winding Caliber RD820SQ Manufacture Roger Dubuis; 166 components; 28,800 vph; 60-hour power reserve.
Functions: hours, minutes.
Case: stainless steel; Ø 42mm; fluted stainless steel bezel; stainless steel crown; antireflective sapphire crystal; stainless steel caseback with sapphire crystal; water resistant to 3 bar.
Dial: anthracite flange; rhodium-plated hour markers; 18K white-gold hands.
Bracelet: stainless steel.

Suggested price: $63,500

EXCALIBUR 42 – AUTOMATIC SKELETON   REF. EX0788

Movement: automatic-winding Caliber RD820SQ Manufacture Roger Dubuis; 166 components; 28,800 vph; 60-hour power reserve.
Functions: hours, minutes.
Case: 18K pink gold; Ø 42mm; fluted 18K pink-gold bezel; 18K pink-gold crown; antireflective sapphire crystal; 18K pink-gold caseback with sapphire crystal; water resistant to 3 bar.
Dial: anthracite flange; gold-plated hour markers; 18K pink-gold hands.
Bracelet: 18K pink gold.

Suggested price: $89,100

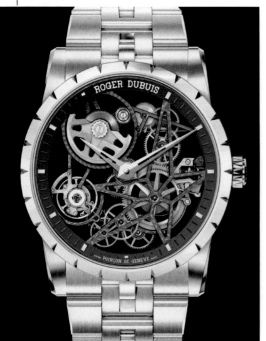
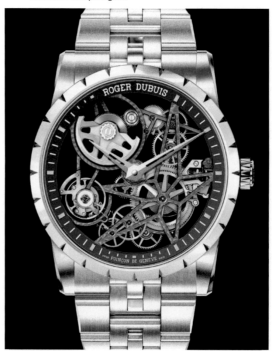

ROGER DUBUIS

EXCALIBUR 36 ASTRAL SKELETON – FLYING TOURBILLON   REF. EX0787

Movement: manual-winding Caliber RD510SQ Manufacture Roger Dubuis; 179 components; 21,600 vph; 60-hour power reserve.
Functions: hours, minutes; single flying tourbillon at 7.
Case: 18K pink gold; Ø 36mm; fluted 18K pink-gold bezel set with 48 white diamonds (1.00 ct); 18K pink-gold crown; antireflective sapphire crystal; 18K pink-gold caseback with sapphire crystal.
Dial: pink-gold-plated flange set with 10 round diamonds (0.15ct); 18K pink-gold hands.
Bracelet: 18K pink gold; quick release system.
Suggested price: $154,000

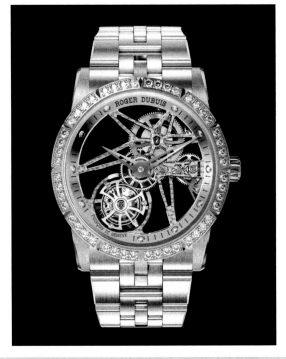

EXCALIBUR 36   REF. EX0588

Movement: automatic-winding Caliber RD830; 183 components; 28,800 vph; 48-hour power reserve.
Functions: hours, minutes; small seconds; date at 6.
Case: 18K pink gold; Ø 36mm; 18K pink-gold crown; fluted 18K pink-gold bezel with 48 round-cut white diamonds (1.00ct); antireflective sapphire crystal; 18K pink-gold caseback with sapphire crystal; water resistant to 5 bar.
Dial: extra white mother-of-pearl; golden Roman numerals.
Strap: purple alligator leather.
Suggested price: $31,200

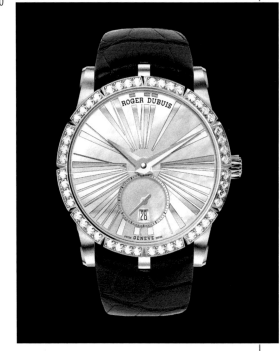

VELVET   REF. VE0069

Movement: automatic-winding Caliber RD830; 183 components; 28,800 vph; 48-hour power reserve.
Functions: hours, minutes.
Case: 18K pink gold; Ø 36mm; 18K pink-gold crown; fluted 18K pink-gold bezel with 64 round-cut white diamonds (0.24 ct); antireflective sapphire crystal; 18K pink-gold caseback with sapphire crystal; water resistant to 3 bar.
Dial: white lacquer; black Roman numerals; 18K pink-gold applied numerals at 6 and 12; 18K pink-gold dauphine hands.
Strap: white alligator leather.
Suggested price: $26,100

VELVET CAVIAR   REF. VE0078

Movement: automatic-winding Caliber RD830; 183 components; 28,800 vph; 48-hour power reserve.
Functions: hours, minutes.
Case: 18K pink gold; Ø 36mm; set with 40 round-cut white diamonds (0.65 ct); 18K pink-gold crown; fluted 18K pink-gold bezel set with 46 round-cut white diamonds (1.00ct); antireflective sapphire crystal; 18K pink-gold caseback with sapphire crystal; water resistant to 5 bar.
Dial: extra white mother-of-pearl; white Roman numerals; 18K pink-gold applied Roman numerals at 6 and 12; 18K pink-gold dauphine hands.
Strap: black crystal fabric; internal rubber lining.
Suggested price: $41,200

ROLEX

ANYWHERE, EVERYWHERE

A titan of the watch industry, Rolex **NEVER FAILS TO INNOVATE, DEVELOPING EVERY MODEL INTO AN EXEMPLARY VISION OF RELIABILITY, PRECISION, AND SOLID MANUFACTURING.** Two new models express this vision. Like all Rolex watches, they bear the Superlative Chronometer certification, a brand-specific standard that exceeds those set by other organizations and affirms, among other attributes, that each model is precise to within -2/+2 seconds per day. The Superlative Chronometer certification also comes with an international five-year guarantee.

A heavy-duty diving watch with impeccable Rolex credentials, the Oyster Perpetual Rolex Deepsea is water resistant to an astounding 3,900m, with every element of the piece designed to enhance its mission. The patented construction of its 44mm Ringlock System case juxtaposes a domed 5.5mm-thick sapphire crystal, a nitrogen-alloyed stainless steel ring within the watch, and a caseback in Oystersteel and grade-5 titanium, which work together to shield the watch against the pressures of the deep. The Triplock winding crown uses three seals and screws down against the case, protecting the watch at what is generally its most vulnerable point. In addition, the Oyster Perpetual Rolex Deepsea is equipped with a helium valve to release excess pressure during decompression. The model's unidirectional rotating bezel, knurled to enhance grip even while wearing gloves, sports a black ceramic Cerachrom insert, marking off 60 minutes of dive time—an essential tool for maintaining diver safety. The movement within, entirely developed and produced by Rolex, is caliber 3235, the result of great strides in watchmaking precision, power reserve, resistance to shocks and magnetism, convenience, and reliability. Benefiting from a new barrel architecture and escapement, caliber 3235 provides a power reserve of approximately 70 hours. The gorgeous dial features a gradual shift in hue from deep blue to black, alluding to the fading, filtered light of a diver's descent. An Oyster bracelet completes the instrument, with an Oysterlock safety clasp that prevents accidental opening and an extension that allows the watch to be worn over a diving suit.

ROLEX

The automatic movement within the GMT-Master II boasts 10 innovations for which Rolex has filed patents, resulting in an exemplary piece of watchmaking from the venerable brand.

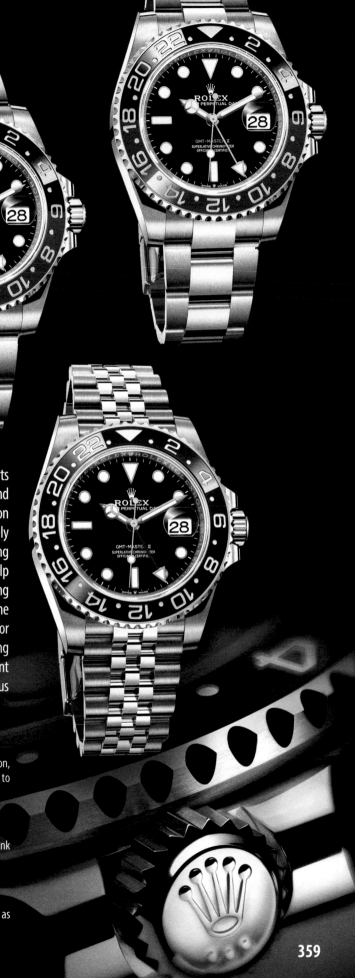

Providing outstanding functionality and satisfying style around the globe, new models in the GMT-Master II line introduce new perspectives in exclusive alloys. The iconic dial layout remains the same: local hours, minutes, and seconds tick around the dial, while a fourth central hand displays home time (or another reference time) on a 24-hour scale around the dial, delineating night and day with bold colors on the bezel. (The date display at 3 o'clock aligns with the local time.) An Oystersteel model complements its monochrome case with a two-tone Cerachrom insert in red and blue ceramic on the bezel, while black and brown Cerachrom inserts grace the bezels of an 18-karat Everose version and an iteration in Everose and Oystersteel (a combination known as Rolesor). These inserts are the culmination of Rolex's research into the production of ceramic elements that are highly resistant to scratches, corrosion, and the effects of ultra-violet rays. Maintaining the emblematic nature of this model, originally developed as a flight tool to help pilots navigate, Rolex is not afraid to continue to evolve the line, redesigning the lugs and sides of the Oyster case and powering the new models with the cutting-edge 3285 caliber. This automatic movement boasts 10 innovations for which Rolex has filed patents, resulting in an exemplary piece of watchmaking from the venerable brand. To name just one, Rolex's Chronenergy escapement is both highly dependable and energy efficient, contributing to its generous power reserve of approximately 70 hours.

▲ **GMT-MASTER II IN EVEROSE AND ROLESOR** (*top left and right*)
The bidirectional rotating bezel of the GMT-Master II is the aesthetic signature of the collection, with a newly developed rich brown ceramic hue joining the black Cerachrom insert to distinguish between day and night on a 24-hour scale.

▲ **GMT-MASTER II IN OYSTERSTEEL** (*bottom*)
In a nod to the brand's iconic history, the GMT-Master II in Oystersteel is mounted on a five-link Jubilee bracelet, originally introduced in 1945 for the Oyster Perpetual Datejust.

◀ **OYSTER PERPETUAL ROLEX DEEPSEA**
The redesigned case of the new Oyster Perpetual Rolex Deepsea features new lugs and sides, as well as a broader bracelet which, like the case, is crafted in Oystersteel.

ROLEX

OYSTER PERPETUAL GMT-MASTER II   REF. 126710 BLRO

Movement: automatic-winding Manufacture Rolex 3285 caliber; 70-hour power reserve; 31 jewels; 28,800 vph; paramagnetic blue Parachrom hairspring with Rolex overcoil; large balance wheel with variable inertia; COSC-certified chronometer.
Functions: hours, minutes, seconds; date at 3; GMT.
Case: Oystersteel; Ø 40mm; bidirectional rotating 24-hour graduated bezel with two-color Cerachrom insert in red and blue ceramic; screw-down winding crown, Triplock triple waterproofness system; scratch-resistant sapphire crystal; screw-down caseback with fine fluting; water resistant to 10atm.

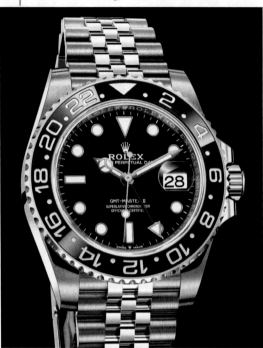

Dial: black lacquer; highly legible Chromalight hour marker appliqués in 18K white gold; 18K white-gold Chromalight hands; red 24-hour hand.
Bracelet: Oystersteel; five-piece solid links; Oysterlock folding safety clasp.
Price: available upon request.

OYSTER PERPETUAL GMT-MASTER II   REF. 126711 CHNR

Movement: automatic-winding Manufacture Rolex 3285 caliber; 70-hour power reserve; 31 jewels; 28,800 vph; paramagnetic blue Parachrom hairspring with Rolex overcoil; large balance wheel with variable inertia; COSC-certified chronometer.
Functions: hours, minutes, seconds; date at 3; GMT.
Case: Oystersteel and 18K Everose gold; Ø 40mm; bidirectional rotating 24-hour graduated bezel with two-color Cerachrom insert in brown and black ceramic; screw-down winding crown, Triplock triple waterproofness system; scratch-resistant sapphire crystal; screw-down caseback with fine fluting; water resistant to 10atm.

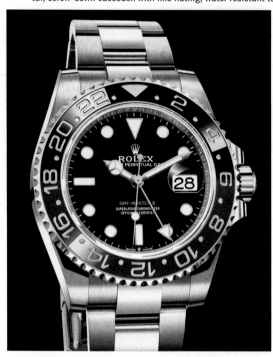

Dial: black lacquer; highly legible Chromalight hour marker appliqués in 18K pink gold; 18K pink-gold Chromalight hands.
Bracelet: Oystersteel and 18K Everose gold; Oysterlock folding safety clasp.
Price: available upon request.

OYSTER PERPETUAL GMT-MASTER II   REF. 126715 CHNR

Movement: automatic-winding Manufacture Rolex 3285 caliber; 70-hour power reserve; 31 jewels; 28,800 vph; paramagnetic blue Parachrom hairspring with Rolex overcoil; large balance wheel with variable inertia; COSC-certified chronometer.
Functions: hours, minutes, seconds; date at 3; GMT.
Case: 18K Everose gold; Ø 40mm; bidirectional rotating 24-hour graduated bezel with two-color Cerachrom insert in brown and black ceramic; screw-down winding crown, Triplock triple waterproofness system; scratch-resistant sapphire crystal; screw-down caseback with fine fluting; water resistant to 10atm.

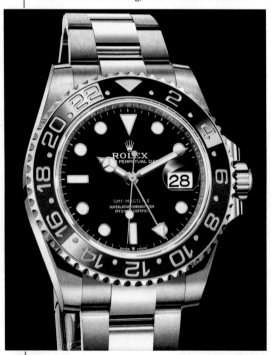

Dial: black lacquer; highly legible Chromalight hour marker appliqués in 18K pink gold; 18K pink-gold Chromalight hands.
Bracelet: 18K Everose gold; Oysterlock folding safety clasp.
Price: available upon request.

COSMOGRAPH DAYTONA   REF. 116595 RBOW

Movement: automatic-winding Manufacture Rolex 4130 caliber; 72-hour power reserve; 44 jewels; 28,800 vph; paramagnetic blue Parachrom hairspring with Rolex overcoil; large balance wheel with variable inertia; COSC-certified chronometer.
Functions: hours, minutes; small seconds at 6; chronograph: 30-minute counter at 3, 12-hour counter at 9, central seconds hand.
Case: 18K Everose gold; Ø 40mm; bezel set with 36 baguette-cut sapphires in a rainbow gradation; screw-down winding crown, Triplock triple waterproofness system; scratch-resistant sapphire crystal; screw-down caseback with fine fluting; water resistant to 10atm.

Dial: black lacquer; 18K pink-gold crystal subdials; 11 baguette-cut, rainbow-colored sapphire hour markers; 18K pink-gold Chromalight hands.
Bracelet: 18K Everose gold; Oysterlock folding safety clasp.
Price: available upon request.

ROLEX

OYSTER PERPETUAL DATEJUST 36 — REF. 126231

Movement: automatic-winding Manufacture Rolex 3235 caliber; 70-hour power reserve; 31 jewels; 28,800 vph; paramagnetic blue Parachrom hairspring with Rolex overcoil; large balance wheel with variable inertia; COSC-certified chronometer.
Functions: hours, minutes, seconds; date at 3.
Case: Oystersteel and 18K Everose gold; Ø 36mm; 18K Everose gold bezel; screw-down winding crown, Twinlock double waterproofness system; scratch-resistant sapphire crystal; screw-down caseback with fine fluting; water resistant to 10atm.
Dial: white lacquer; 18K pink-gold Roman numerals; 18K pink-gold hands.
Bracelet: 18K Everose gold and Oystersteel; folding Oysterclasp.
Price: available upon request.

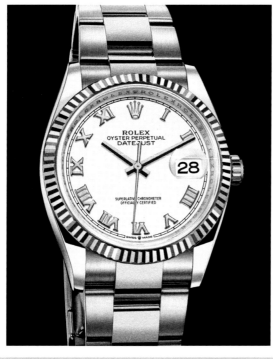

OYSTER PERPETUAL DATEJUST 36 — REF. 126233

Movement: automatic-winding Manufacture Rolex 3235 caliber; 70-hour power reserve; 31 jewels; 28,800 vph; paramagnetic blue Parachrom hairspring with Rolex overcoil; large balance wheel with variable inertia; COSC-certified chronometer.
Functions: hours, minutes, seconds; date at 3.
Case: Oystersteel and 18K yellow gold; Ø 36mm; 18K yellow-gold bezel; screw-down winding crown, Twinlock double waterproofness system; scratch-resistant sapphire crystal; screw-down caseback with fine fluting; water resistant to 10atm.
Dial: champagne; 10 diamonds in 18K yellow-gold settings for hour markers; 18K yellow-gold hands.
Bracelet: 18K yellow gold and Oystersteel; folding Oysterclasp.
Price: available upon request.

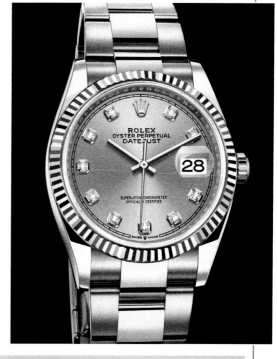

OYSTER PERPETUAL DATEJUST 36 — REF. 126283 RBR

Movement: automatic-winding Manufacture Rolex 3235 caliber; 70-hour power reserve; 31 jewels; 28,800 vph; paramagnetic blue Parachrom hairspring with Rolex overcoil; large balance wheel with variable inertia; COSC-certified chronometer.
Functions: hours, minutes, seconds; date at 3.
Case: Oystersteel and 18K yellow gold; Ø 36mm; 18K yellow-gold bezel set with 52 brilliant-cut diamonds; screw-down winding crown, Twinlock double waterproofness system; scratch-resistant sapphire crystal; screw-down caseback with fine fluting; water resistant to 10atm.
Dial: white mother-of-pearl; 10 diamonds in 18K yellow-gold settings for hour markers; 18K yellow-gold hands.
Bracelet: 18K yellow gold and Oystersteel; folding Oysterclasp.
Price: available upon request.

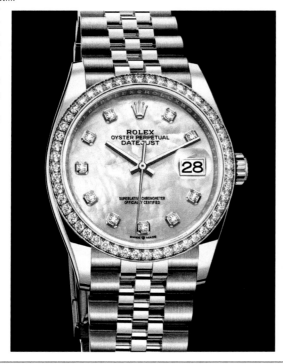

OYSTER PERPETUAL ROLEX DEEPSEA — REF. 126660

Movement: automatic-winding Manufacture Rolex 3235 caliber; 70-hour power reserve; 31 jewels; 28,800 vph; paramagnetic blue Parachrom hairspring with Rolex overcoil; large balance wheel with variable inertia; COSC-certified chronometer.
Functions: hours, minutes, seconds; date at 3.
Case: Oystersteel; Ø 44mm; unidirectional rotating bezel in black ceramic; screw-down winding crown, Triplock triple waterproofness system; domed sapphire crystal; Oystersteel and grade-5 titanium screw-down caseback with fine fluting; water resistant to 390atm.
Dial: gradient from deep blue to pitch black lacquer; 18K white-gold Chromalight hour markers; 18K white-gold Chromalight hands.
Bracelet: Oystersteel; Oysterlock folding safety clasp.
Price: available upon request.

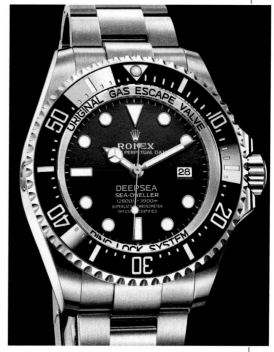

ROLEX

OYSTER PERPETUAL DATEJUST 31 — REF. 278289 RBR

Movement: automatic-winding Manufacture Rolex 2236 caliber; 55-hour power reserve; 31 jewels; 28,800 vph; Syloxi silicon hairspring with patented geometry; balance wheel with variable inertia; COSC-certified chronometer.
Functions: hours, minutes, seconds; date at 3.
Case: 18K white gold; Ø 31mm; bezel set with 46 brilliant-cut diamonds; screw-down winding crown, Twinlock double waterproofness system; scratch-resistant sapphire crystal; screw-down caseback with fine fluting; water resistant to 10atm.
Dial: white mother-of-pearl; 10 diamonds in 18K white-gold settings for hour markers; 18K white-gold hands.
Bracelet: 18K white gold; concealed folding Crownclasp.
Price: available upon request.

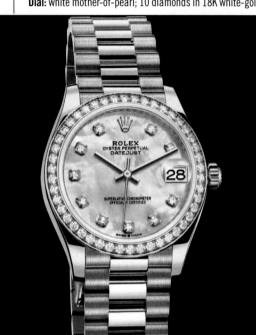

OYSTER PERPETUAL DATEJUST 31 — REF. 278288 RBR

Movement: automatic-winding Manufacture Rolex 2236 caliber; 55-hour power reserve; 31 jewels; 28,800 vph; Syloxi silicon hairspring with patented geometry; balance wheel with variable inertia; COSC-certified chronometer.
Functions: hours, minutes, seconds; date at 3.
Case: 18K yellow gold; Ø 31mm; bezel set with 46 brilliant-cut diamonds; screw-down winding crown, Twinlock double waterproofness system; scratch-resistant sapphire crystal; screw-down caseback with fine fluting; water resistant to 10atm.
Dial: malachite; Roman numeral VI and IX in 18K yellow gold set with 24 diamonds; 18K yellow-gold hands.
Bracelet: 18K yellow gold; concealed folding Crownclasp.
Price: available upon request.

OYSTER PERPETUAL DATEJUST 31 — REF. 278285 RBR

Movement: automatic-winding Manufacture Rolex 2236 caliber; 55-hour power reserve; 31 jewels; 28,800 vph; Syloxi silicon hairspring with patented geometry; balance wheel with variable inertia; COSC-certified chronometer.
Functions: hours, minutes, seconds; date at 3.
Case: 18K Everose gold; Ø 31mm; bezel set with 46 brilliant-cut diamonds; screw-down winding crown, Twinlock double waterproofness system; scratch-resistant sapphire crystal; screw-down caseback with fine fluting; water resistant to 10atm.
Dial: pavé, mother-of-pearl butterfly; 18K pink gold set with 262 diamonds and inlaid with pink mother-of-pearl butterflies; 18K pink-gold hands.
Bracelet: 18K Everose gold; concealed folding Crownclasp.
Price: available upon request.

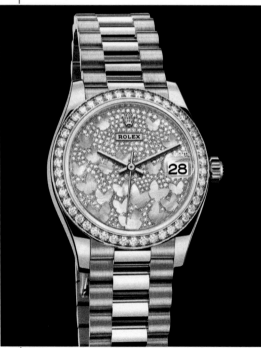

CELLINI MOONPHASE — REF. 50535

Movement: automatic-winding Manufacture Rolex 3195 caliber; 48-hour power reserve; blue Parachrom hairspring; COSC-certified chronometer.
Functions: hours, minutes, seconds; date via central hand; moonphase at 6.
Case: 18K Everose gold; Ø 39mm; fluted double bezel; flared screw-down winding crown with Rolex emblem; domed sapphire crystal; domed screw-down caseback; water resistant to 5atm.
Dial: white.
Strap: tobacco leather; 18K gold buckle.
Price: available upon request.

ROLEX

OYSTER PERPETUAL YACHT-MASTER II — REF. 116680

Movement: automatic-winding Manufacture Rolex 4161 caliber; 72-hour power reserve; blue Parachrom hairspring; COSC-certified chronometer.
Functions: hours, minutes; small seconds at 6; programmable countdown.
Case: 904L steel; Ø 44mm; bidirectional rotatable Ring Command bezel with blue Cerachrom ceramic insert; screw-down winding crown, Triplock triple waterproofness system; scratch-resistant sapphire crystal; water resistant to 10atm.
Dial: white.
Bracelet: 904L steel; Oysterlock folding safety clasp.
Price: available upon request.

OYSTER PERPETUAL YACHT-MASTER 40 — REF. 116695

Movement: automatic-winding Manufacture Rolex 3135 caliber; 48-hour power reserve; 31 jewels; blue Parachrom hairspring with Rolex overcoil; large balance wheel with variable inertia; COSC-certified chronometer.
Functions: hours, minutes, seconds; date at 3.
Case: 18K Everose gold; Ø 40mm; bezel set with 32 sapphires, 8 tsavorites and 1 diamond; screw-down winding crown, Triplock triple waterproofness system; scratch-resistant sapphire crystal; screw-down caseback with Rolex fluting; water resistant to 10atm.
Dial: black; 18K pink-gold hour markers with Chromalight; 18K pink-gold hands with Chromalight.
Price: available upon request.

OYSTER PERPETUAL SEA-DWELLER — REF. 126600

Movement: automatic-winding Manufacture Rolex 3235 caliber; 70-hour power reserve; blue Parachrom hairspring; COSC-certified chronometer.
Functions: hours, minutes, seconds; date at 3.
Case: 904L steel; Ø 43mm; unidirectional rotatable bezel with scratch-resistant Cerachrom ceramic insert; screw-down winding crown, Triplock triple waterproofness system; scratch-resistant sapphire crystal; water resistant to 122atm.
Dial: black.
Bracelet: 904L steel; Oysterlock folding safety clasp.
Price: available upon request.

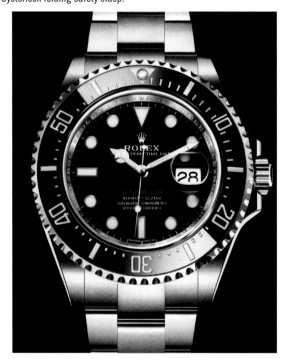

OYSTER PERPETUAL SKY-DWELLER — REF. 326933

Movement: automatic-winding Manufacture Rolex 9001 caliber; 72-hour power reserve; blue Parachrom hairspring; COSC-certified chronometer.
Functions: hours, minutes, seconds; date at 3; month display via 12 apertures around the circumference of the dial; 24-hour display via off-centered disc.
Case: 904L steel and 18K yellow gold; Ø 42mm; fluted bidirectional rotatable Rolex Ring Command bezel; screw-down winding crown, Twinlock double waterproofness system; scratch-resistant sapphire crystal; water resistant to 10atm.
Dial: champagne.
Bracelet: 904L steel and 18K yellow gold; Oysterclasp folding clasp.
Price: available upon request.

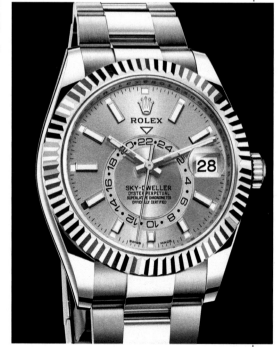

TAGHeuer
SWISS AVANT-GARDE SINCE 1860

TRIED, TESTED, AND TRUE

Using **SOPHISTICATED AESTHETIC CHOICES AND TECHNICAL BRILLIANCE**, TAG Heuer extends its impressive history of collections such as Monaco and Carrera into a future that can hardly be imagined.

The brand recently released a Monaco Gulf special edition, which pays homage to TAG Heuer's racing history by featuring the famous racing colors and logo of Gulf Oil, the company that sponsored Steve McQueen in the film *Le Mans*. The two companies share similar values, proven by their history of innovation, invention and continuing development, while they both have the automotive sport in their DNA. Gulf and TAG Heuer have crossed paths several times, most memorably in 1971, when Steve McQueen wore a Heuer Monaco on his wrist while driving a Porsche 917 that featured the gulf logo. It was the first square and water-resistant automatic chronograph in the history of Swiss watchmaking. The innovative nature of the Monaco lay in its outer appearance (the large square case with straight angles, the metallic blue dial, the crown positioned on the left), its water-resistant technology and in its inner structure. Many of these original features are referenced in the design of the Monaco Calibre 11 Gulf special edition, along with the iconic colors of the Gulf brand. In addition, a light blue and an orange stripe of Gulf's racing colors have been added on the dial. The result is a stunning nod to the heritage of the Swiss watchmaker and the stuff of dreams for automotive racing aficionados.

Heritage and partnership are also celebrated by TAG Heuer's Formula 1 Gulf Special Edition, launched to honor the 50th anniversary of Gulf's first victory in the 24 Hours of Le Mans. TAG Heuer Formula 1 is the ultimate automotive model, synonymous with performance and speed. The watch's round 43mm case frames a petroleum blue dial that also bears the recognizable colors of Gulf: blue and orange. The quartz chronograph features a notched steel bezel and an aluminum ring with a tachometer scale. On the dial, one blue stripe and one orange stripe recall the model's origins, while the caseback is engraved with the Gulf logo. The movement is accurate to a tenth of a second, a feature of paramount importance in a motor racing watch. On the back of the watch can be found Gulf's emblem.

TAG HEUER

Racing red accents on the Carrera Calibre 16 allude to the collection's popularity on the racetrack, including several legendary drivers who have chosen to wear the Carrera on their wrists.

TAG Heuer pays tribute to its sporty roots with the Carrera Calibre 16. Racing red accents gracing a clear chronograph architecture allude unmistakably to the collection's popularity in the realm of motor sport, including several legendary drivers who have chosen to wear the Carrera on their wrists. The Calibre 16 Automatic beats within a 41mm polished steel case, which is water resistant to 100m. Complete with tachometer and date aperture, the Carrera Calibre 16 plays a variation on traditional chronograph layouts, presenting a 30-minute counter at 12 o'clock, a 2-hour counter at 6 o'clock, and small seconds at 9 o'clock. A red central hand ticks off chronograph seconds. Color-coded pushers punctuate the right side of the case, flanking a polished steel crown.

The brand recently paid tribute to its ambassador Ayrton Senna, one of the greatest drivers of all time, with two additions to the Senna special editions collection, this time reinterpreting the iconic TAG Heuer Carrera Chronograph. The Carrera Calibre HEUER02 Chronograph Automatic is available with the iconic bracelet with S-shaped links, as adopted by Ayrton at the height of his achievements. The Carrera Heuer 02 Tourbillon Chronograph Chronometer, available in an individually numbered limited edition of 175 pieces, is mounted on a bi-material rubber/calfskin leather strap, and includes a black matte tachometer with SENNA inscription, displaying the famous Senna symbol and name on the dial, bezel and caseback. The two new watches perfectly reflect Ayrton Senna's sporty and classy spirit.

◀ **TAG HEUER MONACO GULF** (facing page left)
The watchmaker pays tribute to classic racing film *Le Mans* with the Monaco Gulf Special Edition, extending Gulf Oil's racing stripes to the Monaco watch featured in the movie.

◀ **TAG HEUER FORMULA 1 GULF** (facing page right)
Celebrating 50 years since the Gulf's first victory in the 24 Hours of Le Mans, this special edition features the watch's signature petroleum blue dial.

▲ **TAG HEUER CARRERA CALIBRE 16** (top)
Available in two all-business colors, black and rich blue, the Carrera Calibre 16 comes with a matching strap: black leather with red stitching, or steel with H-shaped links.

▲ **TAG HEUER CARRERA HEUER 02T SPECIAL EDITION AYRTON SENNA** (above)
The special edition of the watch that is an homage to the brand's ambassador

TAG HEUER

TAG HEUER CARRERA CHRONOGRAPH GMT — REF. CBG2A1Z.FT6157

Movement: automatic-winding Heuer Calibre 02; 75-hour power reserve.
Functions: hours, minutes; small seconds at 6; date between 4 and 5; chronograph: 30-minute counter at 3, 12-hour counter at 9, central seconds hand; tachometer scale.
Case: steel; Ø 45mm; black and blue ceramic bezel; antireflective sapphire crystal; water resistant to 10atm.
Dial: black skeletonized; rhodium-plated hands and indexes with white SuperLumiNova; "CARRERA HEUER 02 GMT CHRONOGRAPH AUTOMATIC" lettering.
Strap: rubber; steel folding buckle.

Suggested price: $5,950

TAG HEUER CARRERA HEUER 01 MANCHESTER UNITED SPECIAL EDITION — REF. CAR201M.FT6156

Movement: automatic-winding Heuer Calibre 01; Ø 29.3mm; 50-hour power reserve; 39 jewels; 28,800 vph.
Functions: hours, minutes; small seconds at 9; date between 3 and 4; chronograph: 30-minute counter at 12, 12-hour counter at 6, central seconds hand; tachometer scale.
Case: steel; Ø 43mm; antireflective sapphire crystal; steel caseback with special Manchester United logo stamp; water resistant to 10atm.
Dial: brushed and plated with black gold; rhodium-plated hands and indexes with red SuperLumiNova; "CARRERA HEUER 01 CHRONOGRAPH AUTOMATIC" lettering.
Strap: perforated red rubber; steel folding buckle.

Suggested price: $5,750

TAG HEUER CARRERA HEUER 02T SPECIAL EDITION AYRTON SENNA — REF. CAR5A91.FT6162

Movement: automatic-winding Calibre HEUER02T COSC.
Functions: hours, minutes; tourbillon at 6; chronograph: minute counter at 3, hour counter at 9, central seconds counter; COSC-certified chronometer.
Case: black matte ceramic; Ø 45mm; black matte ceramic tachometer fixed bezel; beveled, domed sapphire crystal with anti-reflective treatment; black PVD steel screwed sapphire caseback with SENNA special stamping; water resistant to 100m.
Dial: black opalin; black flange with 60 second/minute scale; rhodium-plate hands and indexes; black SuperLumiNova® on indexes, hour hand, minute hand; rhodium-plated polished TAG Heuer applied logo.
Strap: black calfskin leather/rubber; sand-blasted black titanium folding clasp with double safety pushbuttons.
Note: limited and numbered edition of 175 pieces.

Suggested price: $21,200

TAG HEUER MONACO GULF — REF. CAW211R.FC6401

Movement: automatic-winding Heuer Calibre 11; Ø 30mm; 40-hour power reserve; 59 jewels; 28,800 vph.
Functions: hours, minutes; date at 6; chronograph: 30-minute counter at 9, 60-second counter at 3.
Case: steel; Ø 39mm; domed sapphire crystal; water resistant to 10atm.
Dial: blue sunray brushed; rhodium-plated, facetted hands and indexes with white SuperLumiNova.
Strap: blue calfskin leather; polished steel folding clasp.

Suggested price: $5,900

TAG HEUER

TAG HEUER MONACO CALIBRE 11   REF. CAW211P.FC6356

Movement: automatic-winding Heuer Calibre 11; Ø 30mm; 40-hour power reserve; 59 jewels; 28,800 vph.
Functions: hours, minutes; date at 6; chronograph: 30-minute counter at 9, 60-second counter at 3.
Case: steel; 39x39mm; sapphire crystal; water resistant to 10atm.
Dial: matte blue; faceted indexes and luminescent dots; vintage Heuer logo; "MONACO" & "AUTOMATIC CHRONOGRAPH CALIBRE 11" lettering.
Strap: black perforated calfskin leather; steel folding clasp.
Suggested price: $5,900

TAG HEUER AUTAVIA   REF. CBE2110.BA0687

Movement: automatic-winding, Ø 31mm, thickness: 6.95mm; 75-hour power reserve; 33 jewels; 28,800 vph.
Functions: hours, minutes; small seconds and date at 6; chronograph: 30-minute counter at 3, 12-hour counter at 9, central seconds hand.
Case: steel; Ø 42mm; aluminum black bidirectional rotating bezel; antireflective sapphire crystal; water resistant to 10atm.
Dial: silver sunray brushed; rhodium-plated hands and indexes with white SuperLumiNova; black HEUER printed logo; "AUTAVIA HEUER 02" lettering.
Bracelet: polished steel; polished steel folding clasp.
Suggested price: $6,050

TAG HEUER AQUARACER   REF. WAY101E.FC8222

Movement: quartz.
Functions: hours, minutes, seconds; date at 3.
Case: steel; Ø 43mm; anthracite color aluminum unidirectional rotating bezel; antireflective sapphire crystal; water resistant to 30atm.
Dial: khaki sunray brushed; rhodium-plated, facetted hands and indexes with white SuperLumiNova.
Strap: khaki textile; steel pin buckle.
Suggested price: $1,600

TAG HEUER AQUARACER NIGHT DIVER   REF. WAY108A.FT6141

Movement: quartz.
Functions: hours, minutes, seconds; date at 3.
Case: black PVD grade 2 titanium; Ø 43mm; antireflective sapphire crystal; water resistant to 30atm.
Dial: white textured; black-gold-plated hands and indexes with SuperLumiNova.
Strap: black rubber; brushed black PVD grade 2 titanium folding clasp.
Suggested price: $2,150

TAG HEUER

TAG HEUER AQUARACER LADY AUTOMATIC   REF. WBD2313.BA0740

Movement: automatic-winding Calibre Sellita SW1000; Ø 20.6mm; 40-hour power reserve; 18 jewels; 28,800 vph.
Functions: hours, minutes, seconds; date at 3.
Case: steel; Ø 32mm; steel unidirectional rotating bezel; antireflective sapphire crystal; water resistant to 30atm.
Dial: white mother-of-pearl with stripes decoration; set with 11 diamonds; hands with white SuperLumiNova.
Bracelet: steel; steel folding clasp.

Suggested price: $2,550

TAG HEUER LINK CALIBRE 17 AUTOMATIC   REF. CBC2110.BA0603

Movement: automatic-winding TAG Heuer Calibre 17; Ø 28.6mm; 42-hour power reserve; 37 jewels; 28,800 vph.
Functions: hours, minutes; small seconds at 3; date between 4 and 5; chronograph: 30-minute counter at 9, 12-hour counter at 6, central seconds hand.
Case: steel; Ø 41mm; antireflective sapphire crystal; water resistant to 10atm.
Dial: black; hands and indexes with white SuperLumiNova.
Bracelet: steel; steel double folding clasp.

Suggested price: $4,500

TAG HEUER LADY LINK   REF. WBC1350.BA0600

Movement: quartz.
Functions: hours, minutes, seconds; date at 3.
Case: steel; Ø 32mm; 18K 5N rose-gold bezel; antireflective sapphire crystal; water resistant to 10atm.
Dial: white mother-of-pearl; 5N rose-gold-plated hands and indexes; rose-gold-colored TAG HEUER logo.
Bracelet: steel; steel double folding clasp.
Suggested price: $2,400

TAG HEUER FORMULA 1   REF. WAZ2013.BA0842

Movement: automatic-winding TAG Heuer Calibre 6; Ø 26mm; 38-hour power reserve; 27 jewels; 28,800 vph.
Functions: hours, minutes; small seconds at 6; date at 3.
Case: steel; Ø 43mm; steel unidirectional rotating bezel; sapphire crystal; water resistant to 20atm.
Dial: white; black-gold-plated hands and indexes with white SuperLumiNova; black TAG HEUER logo; "CALIBRE 6 TAG HEUER FORMULA 1" lettering.
Bracelet: brushed steel; brushed steel folding clasp.

Suggested price: $1,750

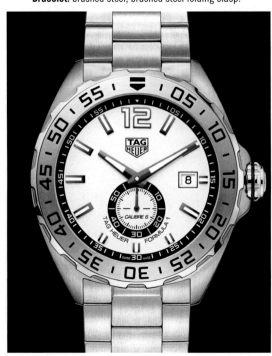

TAG HEUER

TAG HEUER FORMULA 1 ASTON MARTIN SPECIAL EDITION — REF. CAZ101P.FC8245

Movement: quartz.
Functions: hours, minutes, seconds; date at 4; chronograph: 30-minute counter at 9, 60-second counter at 3, 1/10-second counter at 6.
Case: steel; Ø 43mm; matte black aluminum bezel; sapphire crystal; water resistant to 20atm.
Dial: black opaline; black-gold-plated hands and indexes with gray SuperLumiNova; "lime essence" lacquered seconds hand; gray TAG Heuer logo; Aston Martin racing print.
Strap: matte black leather with "lime essence" stitching; steel ardillon buckle with TAG Heuer logo.
Suggested price: $1,650

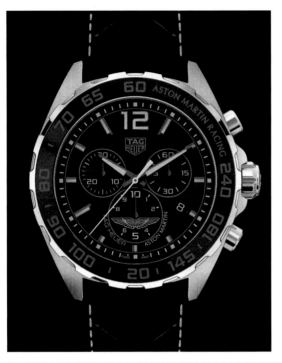

TAG HEUER FORMULA 1 — REF. WBJ1312.FC8231

Movement: quartz.
Functions: hours, minutes, seconds; date at 3.
Case: polished steel; Ø 35mm; sapphire crystal; water resistant to 10atm.
Dial: blue; rhodium-plated hands.
Strap: blue calfskin leather; polished steel pin buckle.
Suggested price: $1,000

TAG HEUER CONNECTED MODULAR 41 — REF. SBF818001.11FT8040

Movement: Ø 41mm, thickness: 13.2mm; powered by Intel Atom Processor.
Functions: hours, minutes, seconds; date and day at 3.
Case: silvered grade 5 titanium; Ø 45mm; antireflective sapphire crystal; caseback engraved with "Intel inside, TAG Heuer, EAC&CE markings"; water resistant to 5atm.
Dial: AMOLED Screen.
Strap: pink calfskin leather with black rubber lining; steel brushed pin buckle.
Suggested price: $1,300

CONNECTED MODULAR 45 – ASTON MARTIN RED BULL RACING SPECIAL EDITION — REF. SBF8A8028

Movement: Ø 45mm, thickness: 13.75mm; powered by Intel Atom Processor.
Functions: hours, minutes, seconds; date at 3; race time via red central hand.
Case: silver-brushed grade 5 titanium; Ø 45mm; scratch-resistant sapphire crystal; water resistant to 5atm.
Dial: AMOLED Screen.
Strap: special dark blue matte calfskin strap with red stitching and black rubber lining; brushed grade 5 titanium folding clasp.
Suggested price: $1,850

VACHERON CONSTANTIN
GENÈVE

BACK TO THE FUTURE

Recalling the innovative spirit of mid-20th-century horology, Vacheron Constantin **REVISITS AN EMBLEMATIC MODEL OF THE ERA** with its FIFTYSIX® collection, which develops timeless style into contemporary standards of excellence.

The year 1956 would prove to be a watershed for Vacheron Constantin, which presented the iconic reference 6073 the same year that Elvis Presley first made it big. Combining a continuity of tradition with a fresh take on the staid case shape, the venerable watchmaker pioneered an approach that would come to dominate the watch world. Each lug of the reference 6073 visually alluded to the Maltese cross, a distinctive shape that has come to represent Vacheron Constantin itself.

VACHERON CONSTANTIN

Vacheron Constantin's reference 6073, from the year 1956, was pivotal: it was one of the first automatic-winding models from Vacheron Constantin, lending it enhanced reliability at a time when many watches required manual winding to maintain their accuracy.

Beyond its external characteristics, reference 6073 was pivotal for another reason: it was one of the first models from Vacheron Constantin's manufacture to be fitted with an automatic-winding movement, lending it enhanced reliability at a time when many watches required manual winding to maintain their accuracy. Another sign of the progress of the times: reference 6073 came housed in a water resistant case. With Vacheron Constantin's new FIFTYSIX® collection, the brand returns to this decisive year, incorporating many advancements watchmaking has developed in the intervening decades. All three models in this collection share a series of characteristics that link them across time and space to the original model.

The FIFTYSIX® self-winding model is available in a 40mm case crafted in steel or 18-karat pink gold. Taking certain aesthetic cues from the original reference 6073, the shape of the case reads as timeless yet modern. The accessible layout alternates hour markers and Arabic numerals, arranging them on a dial that uses both opaline and sunburst finishing. The hour and minute hands echo the shape of the baton hour markers, and like the markers, bear luminescent coating for easy nighttime legibility. A date aperture set into the hour track completes the effortlessly straightforward look. Within beats the new Calibre 1326, whose haute horology touches, from Côtes de Genève, circular graining, and snailing, to the 22-karat gold Maltese cross-shaped oscillating weight, are open to admiration through the sapphire crystal caseback. The movement includes a stop seconds function, making it possible to synchronize the watch with any desired outside source. Vacheron Constantin's unmistakable Maltese cross makes another appearance on the steel folding clasp that accompanies the watch's steel iteration—the pink-gold version comes with a pin buckle, and both styles are mounted on an alligator leather strap.

◀ ▶ **FIFTYSIX SELF-WINDING**
Housed in a 40mm steel or 18-karat pink-gold case, the FIFTYSIX® self-winding model presents a vision of simple elegance that is modern while respecting its vintage roots.

VACHERON CONSTANTIN

Adding a few more useful indications to the dial, the FIFTYSIX® Day-Date displays the date on a snailed counter at 3 o'clock, the day of the week on a similar subdial at 9 o'clock, and a power reserve display at 6 o'clock. Even with these enhancements, the timepiece's face is perfectly in line with the other members of the collection, sporting a subtle palette with an opaline finish in the center and sunburst finishing around the periphery of the dial. Gold Arabic numerals, hour markers, and hour and minute hands lend an understated variation in shade for visual intrigue. The 40mm case comes in steel or 18-karat pink gold, with matching dial appliqués and folding clasp (steel) or buckle (pink gold). Driven by the automatic-winding Calibre 2475 SC/2, the FIFTYSIX® Day-Date features a stop seconds device for utterly controllable accuracy. A sapphire crystal caseback reveals the movement and its pitch-perfect blend of Swiss watchmaking tradition and innovation: the 22-karat gold oscillating weight that winds the movement is openworked and fitted with a ceramic ball bearing rotation system, a technique that negates the need for lubrication and reduces wear and tear on the timepiece's beating heart. The Hallmark of Geneva testifies to the strict standards imposed on the timepiece, crafted in the ancestral home of fine watchmaking.

▶ **FIFTYSIX DAY-DATE**
The FIFTYSIX® Day-Date subtly yet clearly delineates its dial's different zones with varying finishes: opaline, sunburst, and snailed zones guide the eye around the dial.

The standout of the FIFTYSIX® collection is the impeccable FIFTYSIX® Complete Calendar. Its case and dial are immediately recognizable to admirers of the line, with the same opaline-sunburst effect and clever alternation of gold Arabic numerals and hour markers, graced with luminescent coating that facilitates instant readability. The automatic movement inside the Calibre 2460 QCL/1, powers one of horology's most beloved complications: the complete calendar. In addition to hours, minutes and central seconds, the FIFTYSIX® Complete Calendar offers day and month via two apertures at 11 and 1 o'clock, date using a gold arrow that extends to a rim of numbers around the edge of the dial, and a deceptively simple, elegant moonphase display at 6 o'clock, which uses 18-karat gold moons upon a blue night sky. The extreme precision of the moonphase indication means that the wearer need only adjust it once every 122 years (in contrast to a standard moonphase, which requires adjustment every three years). The Calibre 2460 QCL/1 inside beats at 28,800 vph, providing approximately 40 hours of power reserve. Graced with the Hallmark of Geneva as a proof of its extraordinarily high quality, the model reveals its movement, comprising 308 components and 27 jewels, through a transparent sapphire crystal caseback.

▼ **FIFTYSIX COMPLETE CALENDAR**

The FIFTYSIX® Complete Calendar executes one of watchmaking's most coveted complications with effortless grace, presenting not only day, date, and month, but also a precision moonphase display that requires no adjustment for 122 years.

VACHERON CONSTANTIN

PATRIMONY   REF. 81180/000G-9117

Movement: manual-winding 1400 caliber; Ø 20.65mm, thickness: 2.6mm; 40-hour power reserve; 98 components; 20 jewels; Hallmark of Geneva.
Functions: hours, minutes.
Case: 18K white gold; Ø 40mm, thickness: 6.79mm; water resistant to 3atm.
Dial: sand-blasted metal opaline.
Strap: black Mississippiensis alligator leather; 18K white-gold ardillon buckle.
Suggested price: $19,700

PATRIMONY MOON PHASE AND RETROGRADE DATE   REF. 4010U/000G-B330

Movement: automatic-winding 2460 R31L caliber; Ø 27.2mm, thickness: 5.4mm; 40-hour power reserve 275 components; 27 jewels; 28,800 vph; Hallmark of Geneva.
Functions: hours, minutes; retrograde date from 9 to 3; moonphase and age of moon at 6.
Case: 18K white gold; Ø 42.5mm, thickness: 9.7mm; sapphire crystal caseback; water resistant to 3atm.
Dial: silvered opaline; 18K white-gold moon disc; 18K white-gold hour markers.
Strap: black Mississippiensis alligator leather; 18K white-gold buckle.
Suggested price: $43,700

PATRIMONY PERPETUAL CALENDAR   REF. 43175/000R-9687

Movement: automatic-winding 1120QP caliber; Ø 29.6mm, thickness: 4.05mm; 40-hour power reserve; 276 components; 36 jewels; Hallmark of Geneva.
Functions: hours, minutes; perpetual calendar: date at 3, day at 9, month and leap year at 12; moonphase at 6.
Case: 18K 5N pink gold; Ø 41mm, thickness: 8.96mm; sapphire crystal caseback; water resistant to 3atm.
Dial: metal opaline.
Strap: dark brown Mississippiensis alligator leather; 18K 5N pink-gold deployant buckle.
Suggested price: $83,000

PATRIMONY SMALL MODEL   REF. 4100U/110R-B180

Movement: automatic-winding 2450Q6 caliber; Ø 26.2mm, thickness: 3.6mm; 40-hour power reserve; 196 components; 27 jewels; Hallmark of Geneva.
Functions: hours, minutes, seconds; date at 6.
Case: 18K 5N pink gold; Ø 36mm, thickness: 8.1mm; sapphire crystal caseback; water resistant to 3atm.
Dial: metal opaline.
Bracelet: 18K 5N pink gold.
Suggested price: $43,100

VACHERON CONSTANTIN

TRADITIONNELLE   REF. 87172/000R-B403

Movement: automatic-winding 2455 caliber; Ø 26.2mm, thickness: 3.6mm; 40-hour power reserve; 194 components; 27 jewels; 28,800 vph; Hallmark of Geneva.
Functions: hours, minutes; small seconds at 9; date at 3.
Case: 18K 5N pink gold; Ø 38mm, thickness: 8.02mm; sapphire crystal caseback; water resistant to 3atm.
Dial: slate-colored opaline; 18K 5N pink-gold applied hour markers.
Strap: black Mississippiensis alligator leather; 18K 5N pink-gold buckle.
Suggested price: $27,000

TRADITIONNELLE MOON PHASE AND POWER RESERVE SMALL MODEL   REF. 83570/000G-9916

Movement: manual-winding 1410 AS caliber; Ø 26mm, thickness: 4.2mm; 40-hour power reserve; 179 components; 22 jewels; 28,800 vph; Hallmark of Geneva.
Functions: hours, minutes; small seconds at 6; power reserve indicator at 11; moon-phase at 9.
Case: 18K white gold; Ø 36mm, thickness: 9.1mm; bezel and lugs set with 80 round-cut diamonds (1.1 carats); crown set with 1 round-cut diamond; sapphire crystal caseback; water resistant to 3atm.
Dial: white mother-of-pearl.
Strap: gray Mississippiensis alligator leather; 18K white-gold buckle.
Suggested price: $41,400

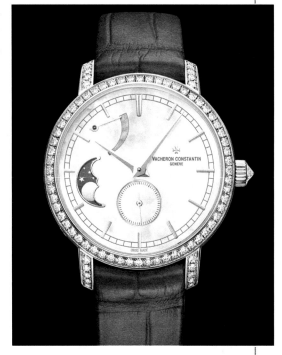

QUAI DE L'ILE   REF. 4500S/000A-B196

Movement: automatic-winding 5100/1 caliber; Ø 30.6mm, thickness: 4.7mm; 60-hour power reserve; 172 components; 37 jewels; 28,800 vph; Hallmark of Geneva.
Functions: hours, minutes, seconds; date via arrow around the dial.
Case: stainless steel; Ø 50.26mm, thickness: 11.75mm; sapphire crystal caseback; water resistant to 3atm.
Dial: black opaline; white luminescent Arabic numerals and indexes; 18K white-gold hour and minute hands with white luminescent material.
Strap: brown Mississippiensis alligator leather; stainless steel folding clasp; delivered with a second black rubber strap.
Suggested price: $16,400

MALTE TOURBILLON OPENWORKED   REF. 30135/000P-9842

Movement: manual-winding 2790SQ caliber; Ø 29.3mm, thickness: 6.1mm; 45-hour power reserve; 246 components; 27 jewels; Hallmark of Geneva.
Functions: hours, minutes; small seconds on tourbillon at 6; date via hand at 2; power reserve indicator at 10.
Case: platinum 950; 38x48.24mm, thickness: 12.73mm; sapphire crystal caseback; water resistant to 3atm.
Dial: sand-blasted skeletonized.
Strap: black Mississippiensis alligator leather; platinum 950 deployant buckle.
Suggested price: $229,000

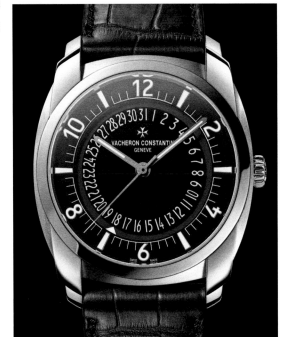

VACHERON CONSTANTIN

OVERSEAS DUAL TIME — REF. 7900V/110A-B333

Movement: automatic-winding 5110 DT caliber; Ø 30.6mm, thickness: 6mm; 60-hour power reserve; 234 components; 37 jewels; 28,800 vph; 22K gold Overseas oscillating weight; Hallmark of Geneva.
Functions: hours, minutes, seconds; second time zone; date at 6; day/night indicator at 9.
Case: stainless steel; Ø 41mm, thickness: 12.8mm; sapphire crystal caseback; water resistant to 15atm.
Dial: silver-toned lacquered; 18K white-gold hour markers and hands with white luminescent material.
Bracelet: stainless steel; delivered with a second black Mississippiensis alligator leather strap; delivered with a third black rubber strap.
Suggested price: $25,400

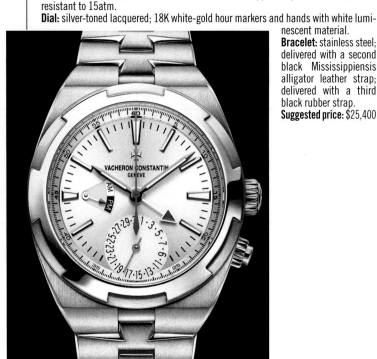

OVERSEAS SMALL MODEL — REF. 2300V/100A-B078

Movement: automatic-winding 5300 caliber; Ø 22.6mm, thickness: 4mm; 44-hour power reserve; 128 components; 31 jewels; 28,800 vph; 22K gold Overseas oscillating weight; Hallmark of Geneva.
Functions: hours, minutes; small seconds at 9.
Case: stainless steel; Ø 37mm, thickness: 11.13mm; sapphire crystal caseback; water resistant to 15atm.
Dial: translucent rosy beige; 18K gold hour markers and hands with white luminescent material.
Bracelet: stainless steel; delivered with a second rosy beige Mississippiensis alligator leather strap; delivered with a third rosy beige rubber strap.
Suggested price: $19,900

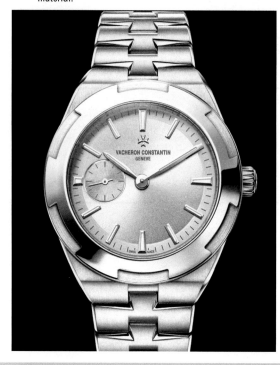

OVERSEAS — REF. 4500V/000R-B127

Movement: automatic-winding 5100 caliber; Ø 30.6mm, thickness: 4.7mm; 60-hour power reserve; 172 components; 37 jewels; 28,800 vph; 22K gold Overseas oscillating weight; Hallmark of Geneva.
Functions: hours, minutes, seconds; date at 3.
Case: 18K 5N pink gold; Ø 41mm, thickness: 11mm; screw-down crown; sapphire crystal caseback; water resistant to 15atm.
Dial: translucent silver-toned lacquered; 18K gold hour markers and hands with white luminescent material.
Strap: brown Mississippiensis alligator leather; delivered with a second brown rubber strap.
Suggested price: $37,200

OVERSEAS CHRONOGRAPH — REF. 5500V/110A-B148

Movement: automatic-winding 5200 caliber; Ø 30.6mm, thickness: 6.6mm; 52-hour power reserve; 263 components; 54 jewels; 28,800 vph; 22K gold Overseas oscillating weight; Hallmark of Geneva.
Functions: hours, minutes; small seconds at 9; date between 4 and 5; chronograph: 12-hour counter at 6, 30-minute counter at 3, central seconds hand.
Case: stainless steel; Ø 42.5mm, thickness: 13.7mm; sapphire crystal caseback; water resistant to 15atm.
Dial: blue lacquered; 18K gold hour markers and hands with white luminescent material.
Bracelet: stainless steel; delivered with a second blue Mississippiensis alligator leather; delivered with a third blue rubber strap.
Suggested price: $30,200

VACHERON CONSTANTIN

OVERSEAS WORLD TIME   REF. 7700V/110A-B172

Movement: automatic-winding 2460 WT/1 caliber; Ø 36.6mm, thickness: 7.55mm; 40-hour power reserve; 255 components; 27 jewels; 28,800 vph; 22K gold Overseas oscillating weight; Hallmark of Geneva.
Functions: hours, minutes, seconds; world time indicator; day/night indicator.
Case: stainless steel; Ø 43.5mm, thickness: 12.6mm; screw-down crown; sapphire crystal caseback; water resistant to 15atm.
Dial: map of northern hemisphere; translucent blue lacquered disc with city names; sapphire disc with day/night and 24-hour indicators; blue lacquered ring; 18K gold hands with white luminescent material.
Bracelet: stainless steel; delivered with a second blue Mississippiensis alligator leather strap; delivered with a third blue rubber strap.
Suggested price: $38,800

HISTORIQUES TRIPLE CALENDRIER 1942   REF. 3110V/000A-B425

Movement: manual-winding 4400 QC caliber; Ø 29mm, thickness: 4.6mm; 65-hour power reserve; 225 components; 21 jewels; 28,800 vph; Hallmark of Geneva.
Functions: hours, minutes; small seconds at 6; calendar: date via central hand, day of the week and month at 12.
Case: stainless steel; Ø 40mm, thickness: 10.35mm; sapphire crystal caseback; water resistant to 3atm.
Dial: sunburst satin-finished silvered; black painted Arabic numerals; burgundy painted date-track.
Strap: brown Mississippiensis alligator leather; steel ardillon buckle.
Suggested price: $20,300

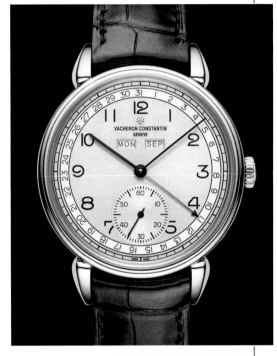

HISTORIQUES AMERICAN 1921 SMALL MODEL   REF. 1100S/000R-B430

Movement: manual-winding 4400 AS caliber; Ø 28.6mm, thickness: 2.8mm; 65-hour power reserve; 127 components; 21 jewels; 28,800 vph; Hallmark of Geneva.
Functions: hours, minutes; small seconds between 4 and 5.
Case: 18K 5N pink gold; Ø 36.5mm, thickness: 7.25mm; sapphire crystal caseback; water resistant to 3atm.
Dial: grained finishing; black painted Arabic numerals.
Strap: red Mississippiensis alligator leather; delivered with a second brown Mississippiensis alligator leather strap; 18K 5N pink-gold buckle.
Suggested price: $30,300

HISTORIQUES ULTRA-FINE 1955   REF. 33155/000P-B169

Movement: manual-winding 1003 caliber; Ø 21.1mm, thickness: 1.64mm; 31-hour power reserve; 117 components; 18 jewels; 18,000 vph; Hallmark of Geneva.
Functions: hours, minutes.
Case: platinum 950; Ø 36mm, thickness: 4.13mm; sapphire crystal caseback; water resistant to 3atm.
Dial: silvered opaline; 18K gold baton hands.
Strap: dark blue Mississippiensis alligator leather; platinum 950 buckle.
Suggested price: $41,900

ZENITH
SWISS WATCH MANUFACTURE SINCE 1865

CHALLENGE ACCEPTED

With a bevy of releases from the Defy collection, Zenith **BOOSTS ITS TECHNICAL BONA FIDES** while giving a few nods to its own groundbreaking path, and a vintage-looking piece pays tribute to an influential subculture.

Of all of Zenith's Defy collection, the Defy El Primero 21 offers the most lingering look at the brand's illustrious history. Named for its world-famous predecessor, an automatic chronograph precise to one-tenth of a second, the Defy El Primero 21 offers an icon for the present century. Incorporating the star-tipped sweep-seconds hand, large, luminescent baton hands, and faceted hour markers of the original El Primero, the Defy El Primero 21 situates these references within a resolutely forward-looking model. Its 44mm lightweight brushed titanium case, water resistant to 100m, frames a blue 30-minute chronograph counter at 3 o'clock and black 60-second counter at 6 o'clock. The true star of the show is the central 100th-of-a-second hand, which dashes around the dial at the mind-bending rate of one full rotation per second, with the 0.01-second increments marked off on the black dial flange. The El Primero 9004 automatic movement drives the time indication at 36,000 vph, while the chronograph derives its astounding precision from a second escapement, which beats at an incredible 360,000 vph—ten times the speed of the original El Primero.

◄ **DEFY EL PRIMERO 21**
With its own escapement, which allows for precision to 1/100th of a second, the chronograph of the Defy El Primero 21 has its own power reserve indication, at 12 o'clock on the dial.

ZENITH

The Defy Classic is resolutely futuristic, with an ultra-light titanium case, a blackened skeleton movement, and a revelation of the activity within.

A star-shaped openworked dial defines the chic, futuristic look of the Defy Classic. Pared down to the essential watchmaking functions, the model bears a date indication at 6 o'clock in addition to the hours, minutes, and seconds. Its color scheme nods to the history of the collection, with a shade of blue that exactly mirrors that of Zenith's El Primero model from 1969. The rest of the watch, however, is resolutely futuristic, with an ultra-light titanium case, a blackened skeleton movement, and a revelation of the activity within. The movement in question is the Elite 670 SK caliber, Zenith's famed base movement reinterpreted to fit the Defy brief: its skeletonized components, extreme visibility, silicon escape-wheel and lever, and black finishing posit mechanical watchmaking as a 21st-century pursuit. Beating at 28,800 vph, it provides a minimum power reserve of 50 hours. Through the sapphire crystal caseback may be seen the special oscillating weight with satin-brushed finish. Framing this fascinating tableau, the 41mm brushed titanium case provides water resistance to 100m, blending well with the titanium bracelet upon which the model is mounted. Another option, a blue alligator leather strap, matches the color of the dial flange.

▲ **DEFY CLASSIC**
A picture of futuristic chic, the Defy Classic nonetheless nods to its pedigree with the color of its dial flange and the star motif of its openworked dial.

ZENITH

The concept behind the tourbillon complication is to negate the effects of gravity on a watch's escapement. Zenith's new Defy Zero G accomplishes the same goal, in a completely novel, even avant-garde manner. Within the manual-winding movement runs Zenith's patented "Gravity Control" gyroscopic module, which obviates the effects of gravity by keeping the horizontal orientation of the regulating organ and balance wheel, an approach inspired by the legendary marine chronometers of old. Upon its introduction in the early part of the 21st century, this module was so imposing that it was placed between two convex sapphire crystals to allow it enough space. Zenith has refined the module so thoroughly that it now takes up just 30% of its previous dimensions, fitting within the confines of the 14.85mm-thick brushed titanium case. The watch is powered by the El Primero 8812S caliber, which beats at 36,000 vph, provides a 50-hour power reserve, and boasts 324 components overall, of which 139 belong to the gyroscopic module. The visual effect of the watch matches the technical sophistication of its signature module. The offset hours and minutes overlap with small seconds at 9 o'clock, leaving a beautiful expanse of airy space within the case. A closer look reveals Zenith's emblematic star motif, emanating out from the gyroscopic module at 6 o'clock on the face. The sapphire crystals front and back reveal the three-dimensional architecture within the watch, contrasting the skeletonized, black rhodium-treated movement with the lighter shade of the star on the dial.

A charitable corporate citizen as well as an horological innovator, Zenith is a proud supporter of The Movember Foundation, an organization dedicated to supporting prostate cancer research and men's mental health. Every fall, over 100,000 people participate in the Distinguished Gentleman's Ride (DGR), driving vintage and classic motorcycles through the streets of 648 cities in over 100 countries to raise awareness of and funds for these important public health issues. Team Zenith is one of the largest teams to ride in the DGR, as well as one of the event's top fundraisers. To celebrate this worthy cause and unique event, Zenith unveiled the Pilot Type 20 Chronograph Ton-Up Black. The watch's name stems from the Café Racers of 1950s Britain, who embraced motorcycling as an expression of their identity, and performed endless modifications on their machines to hit 100 miles per hour (also known as a "ton"). Zenith thus pays tribute to a vintage subculture that emphasized non-conformism and a relentless drive to perfect one's own mechanical stats. Pairing a classic look with sporty mechanics, this model draws its power from the El Primero 4069 column-wheel chronograph, which beats at 36,000 vph and provides a power reserve of 50 hours. The all-black look of this model lends a dramatic flair to its retro vibe, from its matte dial to its 45mm aged stainless steel case, with luminescent coating on the hands and Arabic numeral hour markers enhancing the model's instant legibility. The watch's caseback features Café Racer Spirit engravings, completing a perfect homage.

▶ **DEFY ZERO G**
With a diameter of 44mm, thickness of 14.85mm, and water resistance to 100m, the titanium case of the Defy Zero G protects a "Gravity Control" module that has been shrunk to just 30% of its original size.

ZENITH

▶ **PILOT TYPE 20 CHRONOGRAPH TON-UP BLACK**

Channeling the intensity of British biker culture from the mid-20th century, the Pilot Type 20 Chronograph Ton-Up Black offers a vintage look with an fashion-forward aesthetic, powered by 21st-century mechanics.

ZENITH

DEFY CLASSIC — REF. 95.9000.670/51.M9000

Movement: automatic-winding Elite 610 SK caliber; 50-hour power reserve; 27 jewels; 28,800 vph.
Functions: hours, minutes, seconds; date at 3.
Case: titanium; Ø 41mm, thickness: 10.75mm; antireflective sapphire crystal; sapphire crystal caseback; water resistant to 10 atm.
Dial: blue.
Bracelet: titanium.
Suggested price: $6,900

DEFY CLASSIC — REF. 95.9000.670/78.R782

Movement: automatic-winding Elite 670SK caliber; 50-hour power reserve; 27 jewels; 28,800 vph.
Functions: hours, minutes, seconds; date at 6.
Case: titanium; Ø 41mm, thickness: 10.75mm; antireflective sapphire crystal; sapphire crystal caseback; water resistant to 10 atm.
Dial: skeletonized.
Strap: black rubber.
Suggested price: $6,500

DEFY EL PRIMERO 21 — REF. 95.9000.9004/78.R582

Movement: automatic-winding El Primero 9004 caliber; 50-hour power reserve; 53 jewels; 36,000 vph.
Functions: hours, minutes; small seconds at 9; chronograph: 30-minute counter at 3, 60-second counter at 6; power reserve indicator at 12.
Case: titanium; Ø 44mm, thickness: 14.50mm; antireflective sapphire crystal; sapphire crystal caseback; water resistant to 10 atm.
Dial: skeletonized.
Strap: black rubber coated with black alligator leather coating.
Suggested price: $11,200

DEFY EL PRIMERO 21 — REF. 49.9000.9004/78.M9000

Movement: automatic-winding El Primero 9004 caliber; 50-hour power reserve; 53 jewels; 36,000 vph.
Functions: hours, minutes, small seconds at 9; chronograph: 30-minute counter at 3, 60-second counter at 6; power reserve indicator at 12.
Case: ceramic; Ø 44mm, thickness: 14.50mm; antireflective sapphire crystal; sapphire crystal caseback; water resistant to 10 atm.
Dial: skeletonized.
Bracelet: black ceramic.
Suggested price: $15,100

ZENITH

PILOT CRONOMETRO TIPO CP-2 FLYBACK   REF. 29.2240.405/18.C801

Movement: automatic-winding El Primero 405B caliber; Ø 30mm, thickness: 6.6mm; 50-hour power reserve; 254 components; 31 jewels; 36,000 vph.
Functions: hours, minutes, small seconds at 9; chronograph: 30-minute counter at 3, flyback.
Case: bronze; Ø 43mm, thickness: 12.85mm; antireflective sapphire crystal; sapphire crystal caseback; water resistant to 10 atm.
Dial: bronze grained.
Strap: brown oily nubuck leather; protective rubber lining; titanium pin buckle.
Suggested price: $7,700

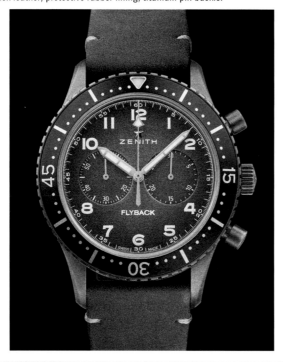

PILOT TYPE 20 EXTRA SPECIAL   REF. 29.1940.679/21.C800

Movement: automatic-winding Elite 679 caliber; 50-hour power reserve; 27 jewels; 28,800 vph.
Functions: hours, minutes, seconds.
Case: bronze; Ø 40mm, thickness: 12.95mm; antireflective sapphire crystal; water resistant to 10 atm.
Dial: black sunray.
Strap: green nubuck leather.
Suggested price: $5,700

PILOT TYPE 20 CHRONOGRAPH TON-UP   REF. 11.2432.4069/21.C900

Movement: automatic-winding El Primero Caliber 4069; 50-hour power reserve; 35 jewels; 36,000 vph.
Functions: hours, minutes, small seconds at 9; chronograph: 30-minute counter at 3.
Case: aged stainless steel; Ø 45mm, thickness: 14.25mm; antireflective sapphire crystal; water resistant to 10 atm.
Dial: matte black.
Strap: black nubuck leather.
Suggested price: $7,100

PILOT TYPE 20 CHRONOGRAPH EXTRA SPECIAL   REF. 29.2430.4069/57.C808

Movement: automatic-winding El Primero 4069 caliber; 50-hour power reserve; 35 jewels; 36,000 vph.
Functions: hours, minutes, small seconds at 9; chronograph; 30-minute counter at 3.
Case: bronze; Ø 45mm, thickness: 14.25mm; antireflective sapphire crystal; titanium caseback; water resistant to 10 atm.
Dial: matte blue.
Strap: blue nubuck leather.
Suggested price: $7,100

[Brand Directory]

A. LANGE & SÖHNE
Altenberger Strasse 15
01768 Glashütte
Germany
Tel: 49 35053440
USA: 800 408 8147
Arije: 33 1 47 20 50 50

AUDEMARS PIGUET
1348 Le Brassus
Switzerland
Tel: 41 21 845 14 00
USA: 212 758 8400
Arije: 33 1 47 20 50 50

BLANCPAIN
6 Chemin de l'Etang
1094 Paudex Switzerland
Tel: 41 21 796 36 36
USA: 877 520 1735
Arije: 33 1 47 20 50 50

BREGUET
23 Place de la Tour
1344 L'Abbaye
Switzerland
Tel: 41 21 841 90 90
USA: 866 458 7488
Arije: 33 1 47 20 50 50

BREITLING
Schlachthausstrasse 2
2540 Grenche
Switzerland
Tel: 41 32 925 95 25
USA: 203 762 1180
Arije: 33 1 47 20 50 50

BVLGARI
34 Rue de Monruz
2000 Neuchâtel
Switzerland
Tel: 41 32 722 78 78
USA: 212 315 9700
France: 33 1 53239090
Arije: 33 1 47 20 50 50

CARL F. BUCHERER
1805 South Metro Parkway
Dayton, OH 45459
USA: 800 395 4306

CARTIER SA
8 Boulevard James-Fazy
1201 Geneva
Switzerland
Tel: 41 22 721 24 00
USA: 212 446 3400
Arije: 33 1 47 20 50 50

CHAUMET
12 Place Vendôme
75001 Paris
France
Tel: 33 1 44 77 24 00
Arije: 33 1 47 20 50 50

CHOPARD
8 Rue de Veyrot
1217 Meyrin
Geneva 2
Switzerland
Tel: 41 22 719 3131
USA: 212 821 0300
Arije: 33 1 47 20 50 50

de GRISOGONO
39 Chemin du Champs des Filles
Bâtiment E
1228 Plan-les-Ouates
Switzerland
Tel: 41 22 817 81 00
USA: 212 439 4220
Arije: 33 1 47 20 50 50

DIOR HORLOGERIE
44 Rue François 1er
75008 Paris
France
Tel: 33 1 40 73 59 84
USA: 212 931 2700

FRANCK MULLER
Karia Luxury Watch Distribution
18, rue Volney
75002 Paris
France: 33 1 84 17 84 80
USA: 212 463 8898
Arije: 33 1 47 20 50 50

GIRARD-PERREGAUX
1 Place Girardet
2301 La Chaux-de-Fonds
Switzerland
Tel: 41 32 911 33 33
USA: 201 804 1904 ext. 2224
Arije: 33 1 47 20 50 50

GUY ELLIA
7 rue Lincoln
75008 Paris
France
Tel: 33 1 53 30 25 25
USA: 212 888 0505
Arije: 33 1 47 20 50 50

HUBLOT
33 Chemin de la Vuarpillière
1260 Nyon 2
Switzerland
Tel: 41 22 990 90 00
USA: 800 536 0636
Arije: 33 1 47 20 50 50

IWC
Baumgartenstrasse 15
8201 Schaffhausen
Switzerland
Tel: 41 52 635 65 65
USA: 1 800 492 6755
Arije: 33 1 47 20 50 50

JACOB & CO.
1 Chemin de Plein-Vent
1228 Arare
Switzerland
Tel: 41 22 310 69 62
USA: 212 719 5887
Arije: 33 1 47 20 50 50

JAQUET DROZ SA
2 Allée du Tourbillon
2300 La Chaux-de-Fonds
Switzerland
Tel: 41 32 924 28 88
USA: 201 271 1400
Arije: 33 1 47 20 50 50

LONGINES
Saint-Imier
2610 Switzerland
Tel: 41 32 942 54 25
USA: 201 271 1400

LOUIS VUITTON
101 av. des Champs-Élysées
75008 Paris
France
Tel: 33 1 53 57 52 00
USA: 866 VUITTON

OMEGA
Rue Jakob Staempfli 96
2500 Bienne 4
Switzerland
Tel: 41 32 343 9211
USA: 201 271 1400
Arije: 33 1 47 20 50 50

PATEK PHILIPPE
141 Chemin du Pont du Centenaire
1228 Plan-les-Ouates
Switzerland
Tel: 41 22 884 20 20
USA: 212 218 1240

PIAGET
37 Chemin du Champ-des-Filles
1228 Plan-les-Ouates
Switzerland
Tel: 41 22 884 48 44
USA: 212 891 2440
Arije: 33 1 47 20 50 50

RICHARD MILLE
11 rue du Jura
2345 Les Breuleux Jura
Switzerland
Tel: 41 32 959 43 53
France: 33 2 99 49 19 00
USA: 310 205 5555

ROGER DUBUIS
Rue André de Garrini
1217 Meyrin 2 Geneva
Switzerland
Tel: 41 22 808 48 88
USA: 212 651 3773
Arije: 33 1 47 20 50 50

ROLEX
3-5-7 Rue François Dussaud
1211 Geneva 26
Switzerland
Tel: 41 22 302 22 00
USA: 212 758 7700
Arije: 33 1 47 20 50 50

TAG HEUER
6A Louis-Joseph Chevrolet
2300 La Chaux-de-Fonds
Switzerland
Tel: 41 32 919 80 00
USA: 973 467 1890
Arije: 33 1 47 20 50 50

VACHERON CONSTANTIN
10 Chemin du Tourbillon
1228 Plan-les-Ouates
Switzerland
Tel: 41 22 930 20 05
USA: 212 713 0707
Arije: 33 1 47 20 50 50

ZENITH
2400 Le Locle
Switzerland
Tel: 41 32 930 62 62
USA: 973 467 1890
Arije: 33 1 47 20 50 50